**Willis's Elements of Quantity Surveying**

# Willis's Elements of Quantity Surveying

Fourteenth Edition

Roy Hills
Sandra Lee

**WILEY** Blackwell

This fourteenth edition first published 2024
© 2024 John Wiley & Sons Ltd

*Edition History*
John Wiley & Sons Ltd (13e, 2020)

*Registered Offices*
John Wiley & Sons, Inc., 111 River Street, Hoboken, NJ 07030, USA
John Wiley & Sons Ltd, The Atrium, Southern Gate, Chichester, West Sussex, PO19 8SQ, UK

For details of our global editorial offices, customer services, and more information about Wiley products visit us at www.wiley.com.

Wiley also publishes its books in a variety of electronic formats and by print-on-demand. Some content that appears in standard print versions of this book may not be available in other formats.

**Library of Congress Cataloging-in-Publication Data Applied for:**

Paperback ISBN: 9781394177820

Cover Design: Wiley
Cover Images: © jailce/Adobe Stock Photos, xu wu/Getty Images, Paul Bradbury/Getty Images, Photo_Concepts/Getty Images, Richard Newstead/Getty Images

Set in 9.5/12.5pt STIXTwoText by Straive, Chennai, India
Printed and bound by CPI Group (UK) Ltd, Croydon, CR0 4YY

C9781394177820_010224

# Contents

# Abbreviations

The continued publication of this text would not be possible without the input of the previous authors William Trench and Andrew... The efforts of Ruth Watson are gratefully acknowledged in making the necessary alterations to the drawings.

Abbreviations as before.
Abbreviations before described.

| | | |
|---|---|---|
| BCIS | Building Cost Information Service | |
| BS | British Standard | |
| CAWS | Common Arrangement of Work Sections | |
| c/s | centres | |
| ccu | cooker control unit | |
| ded | deduct | |
| dia | diameter | |
| d.p.c | damp-proof course | |
| d.p.m | damp-proof membrane | |
| EDI | electronic data interchange | |
| e.w.s | earth work support | |
| o/o | out of | |
| hw | hardwood | |
| JCT | Joint Contracts Tribunal | |
| LED | light emitting diode | |
| MC | Measurement... | |

# 1

# Introduction

## The modern quantity surveyor

The training and knowledge of the quantity surveyor have enabled the role of the profession to evolve over time into new areas, and the services provided by the modern quantity surveyor now cover all aspects of procurement, contractual, and project cost management. This holds true whether the quantity surveyor works as a consultant or is employed by a contractor or subcontractor. Whilst the importance of this expanded role cannot be emphasised enough, success in carrying it out stems from the traditional ability of the quantity surveyor to measure and value. It is on the aspect of measurement that this book concentrates.

## The need for measurement

There is a need for measurement of a proposed construction project at various stages from the feasibility stage through to the final account. This could be in order to establish a budget price, give a pre-tender estimate, produce contract documents for pricing, provide a contract tender sum, or evaluate the amount to be paid to a contractor. There are many construction or project management activities that require some form of measurement so that appropriate rates can be applied to the quantities and a price or cost established.

The measurement explained in this book is primarily for the production of a bill of quantities as part of the traditional procurement approach to construction. Other procurement approaches move the need for detailed measurement to later stages of the project cycle and away from activity undertaken by the client's team to that of the contractor's team.

## The need for rules

The need for rules to be followed when undertaking any measurement becomes clear when costs for past projects are analysed and elemental rates or unit rates are calculated and then applied to the quantities for a proposed project. For greater accuracy in pricing, it is

*Willis's Elements of Quantity Surveying*, Fourteenth Edition. Roy Hills and Sandra Lee.
© 2024 John Wiley & Sons Ltd. Published 2024 by John Wiley & Sons Ltd.

# 2
# Detailed Measurement

## Method of analysing cost

It is evident that if a building is divided up into its constituent parts, and the cost of each part can be estimated, an estimate can be compiled for the whole work. It was found in practice that by making a 'schedule' setting out the quantity of each item of work for a project, the labour and material requirements for these could be more readily assessed. This schedule at Royal Institute of British Architects (RIBA) stage 4 can be in the form of a bill of quantities which, when priced by a contractor, provides a tender sum for a project. It must not be forgotten that a traditional bill of quantities only produces an estimate. It is prepared and priced before the erection of the building and gives the contractor's estimated cost. Such an estimated cost, however, under the most commonly used construction contracts, becomes a tender and a definite price for which the contractor agrees to carry out the work as set out in the bill. The bill must, therefore, completely represent the proposed work so that a serious discrepancy between actual and estimated cost does not arise.

## Origin of the bill of quantities

Competitive tendering is one of the basic principles of most classes of business, and if competitors are given comprehensive details of the requirements, it should be fair to all concerned. However, historically when tendering based on drawings and specification, builders found that considerable work was involved in making detailed calculations and measurements to form the basis for a tender. They realised that by getting together and employing one person to make these calculations and measurements for them all, a considerable saving would be made in their overhead charges. They began to arrange for this to be done, each including the surveyor's fee for preparing the bill of quantities in their tender, and the successful competitor paying. Each competing builder was provided with the same bill of quantities which could then be priced in a comparatively short space of time. It was not long before this situation was realised by the architect and employer. Here the employer was paying indirectly for the quantity surveyor through the builder, whereas the surveyor could be used as a consultant if a direct appointment was made. This would give the employer

*Willis's Elements of Quantity Surveying*, Fourteenth Edition. Roy Hills and Sandra Lee.
© 2024 John Wiley & Sons Ltd. Published 2024 by John Wiley & Sons Ltd.

Nevertheless, the contractor's surveyor must be able to check a bill of quantities and measure variations on the basis of that bill. It is therefore essential that the contractor's surveyor should understand how the bill is prepared, and there should be no difficulty in adapting this knowledge to suit the somewhat different requirements when preparing quantities for a contractor's estimate.

## Differences of custom

It must be understood that, as a good deal of the subject matter of this book is concerned with method and procedure, suggestions made must not be taken as invariable rules. Surveyors will have, in many cases, their own customs and methods of working which may differ from those given here, and which may be equally good or, in their view, better. The procedure advocated is put forward as being reasonable and based on practice. Furthermore, all rules must be adapted to suit any particular circumstances of the project to be measured.

# 3

# The Use of the RICS *New Rules of Measurement* (NRM)

## Background

The *New Rules of Measurement* (NRM) Project was arguably one of the most significant developments in quantity surveying practice since the publication of the SMM7 in 1988. The intention of the Royal Institution of Chartered Surveyors (RICS) with this suite of documents is to create a set of common rules that provide a consistent approach to measurement through the various stages of a project, from initial cost estimate to detailed quantification of the construction work. Whilst the NRM is based on UK practice, it is nevertheless intended to have worldwide application.

This chapter therefore looks at the key features of the NRM in order to explain how it relates to the measurement covered in this text.

Historically, surveyors would approach the measurement of approximate quantities in different ways: for example, the area of external walls might be measured over windows and doors (i.e. gross measurement) by one surveyor, whilst another might deduct the area of the windows and doors (i.e. net measurement). The method of measurement would be closely related to the way in which the rates were to be applied. This variation in practice then resulted in an inconsistent approach to early estimates and cost planning, which then led to further problems when cost plans were used instead of bills of quantities as the basis of tender negotiations. Cost plans would then be analysed and used as benchmarks for further cost plans, thus creating an unreliable database and potentially inaccurate estimates. There was also a lack of continuity in cost data between cost plans and bills of quantities, making it almost impossible to reconcile the cost plan with the priced bill of quantities or pre-tender estimate.

### The NRM volumes

The RICS has published a set of documents in three volumes, for the measurement of building work from the early feasibility stage through to completion, handover, and building occupation. The full NRM comprises:

*Willis's Elements of Quantity Surveying,* Fourteenth Edition. Roy Hills and Sandra Lee.
© 2024 John Wiley & Sons Ltd. Published 2024 by John Wiley & Sons Ltd.

NRM1 – Order of cost estimating and cost planning for capital building works, covering:

- Estimating – Royal Institute of British Architects (RIBA) Work Stages 0-1; Office of Government Commerce (OGC) Gateways 1 and 2.
- Cost planning – Elements – RIBA Work Stages 2–4; OGC Gateways 3A, 3B, and 3C.

NRM2 – Detailed measurement for building works, covering:

- RIBA Work Stages 4, 5 and 6; OGC Gateway 3C (detailed measurement for tender documentation).

NRM3 – Order of cost estimating and cost planning for building maintenance works, covering:

- RIBA Work Stage 7; OGC Gateways 4 and 5 (life cycle costing).

## Introduction to NRM2

There is still a need for the preparation of bills of quantities following the rules of a standard method of measurement, where detailed information is provided as the basis for the traditional, fixed-price, lump-sum approach to procurement. NRM2 contains the rules for measurement to be used when preparing bills of quantities, and it is therefore used as the basis for explaining how to measure the examples included in this text. Table 4 gives the various work sections that are included in NRM2, and examples of measurement using these rules are included in Chapters 8–19.

**Table 4**   Extract from Section 3 of NRM2

**Work sections 2–41 comprise the rules of measurement for building components and items. They are as follows:**

| | |
|---|---|
| 2 | Off-site manufactured materials, components, and buildings |
| 3 | Demolitions |
| 4 | Alterations, repairs, and conservation |
| 5 | Excavating and filling |
| 6 | Ground remediation and soil stabilisation |
| 7 | Piling |
| 8 | Underpinning |
| 9 | Diaphragm walls and embedded retaining walls |
| 10 | Crib walls, gabions, and reinforced earth |
| 11 | In-situ concrete works |
| 12 | Precast/composite concrete |
| 13 | Precast concrete |
| 14 | Masonry |
| 15 | Structural metalwork |

**Table 4** (Continued)

**Work sections 2–41 comprise the rules of measurement for building components and items. They are as follows:**

| | |
|---|---|
| 16 | Carpentry |
| 17 | Sheet roof coverings |
| 18 | Tile and slate roof and wall coverings |
| 19 | Waterproofing |
| 20 | Proprietary linings and partitions |
| 21 | Cladding and covering |
| 22 | General joinery |
| 23 | Windows, screens, and lights |
| 24 | Doors, shutters, and hatches |
| 25 | Stairs, walkways, and balustrades |
| 26 | Metalwork |
| 27 | Glazing |
| 28 | Floor, wall, ceiling, and roof finishings |
| 29 | Decoration |
| 30 | Suspended ceilings |
| 31 | Insulation, fire stopping, and fire protection |
| 32 | Furniture, fittings, and equipment |
| 33 | Drainage above ground |
| 34 | Drainage below ground |
| 35 | Site works |
| 36 | Fencing |
| 37 | Soft landscaping |
| 38 | Mechanical services |
| 39 | Electrical services |
| 40 | Transportation |
| 41 | Builder's work in connection with mechanical, electrical, and transportation installations |

NRM2 has a detailed section on how to code bills of quantities, but the link between elemental cost plans and bills in trade order can be seen only if bills are accurately coded when produced to enable sorting between the two different formats. Further information on bill preparation is provided in Chapter 21.

PDF copies of the NRM volumes are available as a free download from the RICS website to encourage all to use the guidance.

# 4

# Setting Down Dimensions

The development of computerised measurement and billing systems, each with its own structure for inputting dimensions and calling up descriptions, has made the more traditional, manual procedures less common nowadays. However, it is only by understanding the basic principles of setting down dimensions and descriptions in traditional form, as detailed here, that one can then apply them to whatever measurement process is adopted. It is also essential to understand the measurement of another surveyor so that work can be checked, and it is only by following the approach to setting down dimensions precisely that this can happen.

## Traditional dimension paper

The dimensions are measured from the drawings by the taker-off, using paper ruled as follows:

| 1 | 2 | 3 | 4 | 1 | 2 | 3 | 4 |
|---|---|---|---|---|---|---|---|
|   |   |   |   |   |   |   |   |

The columns (not, of course, normally numbered) have been numbered here for identification. Column 1 is called the *timesing column*, and its use will be described later. Column 2 is the *dimension column* in which the measurements are set down as taken from the drawings. Column 3 is the *squaring column* in which are set out the calculated volumes, areas, and so on of the measurements in column 2. Column 4 is the *description column* in which is written the description of the work to which the dimensions apply, and on the extreme right-hand side of which (known as *waste*) preliminary calculations and

*Willis's Elements of Quantity Surveying*, Fourteenth Edition. Roy Hills and Sandra Lee.
© 2024 John Wiley & Sons Ltd. Published 2024 by John Wiley & Sons Ltd.

collections are made. There are two sets of columns in the width of a single A4 sheet. No written work should be carried across the central vertical division. There is usually a narrow binding margin (not shown here) on the left of the sheets. The taking-off commences at the top of the description column on the left-hand side and continues down the column until the bottom is reached and then the process continues at the top of the following description column on the right-hand side of the same page. In the examples that follow, an extra column has been included for the purpose of including teaching notes as necessary.

Each dimension sheet should have the name and/or the number of the project written on or, better, stamped on. In addition, the title of the section being measured should be included, followed by a number, starting at 1 for each section. The unique numbering of the dimension sheets allows the surveyor to easily find an item at a later date. The examples measured in the following chapters have been written using only one-half of the sheet with the right-hand side being used for explanation.

## Form of dimensions

Before going any further, it is necessary to understand the dimensions as set down by the taker-off. All dimensions are in one of five forms:

1) Cubic measurements
2) Square measurements
3) Linear measurements
4) Enumerated items
5) Items.

These are expressed in the first three cases by setting down the measurements immediately under each other in the dimension column, with each separate item being divided from the next by a line, for example:

| | | | |
|---|---|---|---|
| | 3.00<br>2.00<br>4.00 | indicating a cubic measurement<br>3.00 m long, 2.00 m wide and<br>4.00 m deep or tall | |
| | 3.00<br>2.00 | indicating a square metre<br>measurement 3.00 m long by<br>2.00 m wide | |
| | 3.00 | indicating a linear measurement<br>3.00 m long | |

Here we can see that the length dimension is written first followed by the width and then the depth or height which removes any ambiguity over which figure relates to each.

An item to be enumerated is usually indicated in one of the following ways:

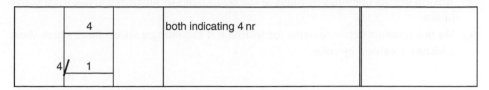

Occasionally, NRM2 requires the use of an item; this is a description without a measured quantity (e.g. testing the drainage system). The description may, if applicable, contain dimensions (e.g. temporary screens). This requirement is indicated as follows:

There is no need to label dimensions *m3*, *m2*, *m*, if a rule is made always to draw a line under each measurement; it is obvious from the number of entries in the measurement under which category it comes. Also, enumerated items may be distinguished from the above by the use of whole numbers only. It is usual to set down the dimensions in the following order:-

1) Horizontal length
2) Horizontal width or breadth
3) Vertical depth or height.

Although the order will not affect the calculations of the cubic or square measurement, it is valuable in tracing measurements later if a consistent order is maintained. As will be explained in this book, an incorrect order in a description may even sometimes mislead an estimator in pricing.

The following is an extract from NRM2 section 3.2.1:

The rules for quantifying building components/items are as follows:

1) *Measurement and billing*:
   a) Measure work net as fixed in position unless otherwise stated.
   b) Net quantity measured shall be deemed to include all additional material required for laps, joints, seams, and the like as well as any waste material.
   c) Curved work shall be measured on the centre line of the material unless otherwise stated.
   d) Dimensions shall be measured to the nearest 10 mm. 5 mm and over shall be regarded as 10 mm and less than 5 mm shall be disregarded.
   e) Except for quantities measured in tonnes (t), quantities shall be given to the nearest whole number. Quantities less than one unit shall be given as one unit. Quantities measured in tonnes (t) shall be given to two decimal places.

2) *Voids*:
   a) Unless otherwise stated, minimum deductions for voids refer only to openings or wants within the boundaries of the measured work.
   b) Always deduct openings or wants at the boundaries of measured areas, irrespective of size.
   c) Do not measure separate items for widths not exceeding a stated limit where these widths are caused by voids.

## Timesing

It often happens that when the taker-off has written the dimension, it is found that there are several items having the same measurements. To indicate that the measurement is to be multiplied, it will be *timesed* as follows:

| | | | |
|---|---|---|---|
| 3/ | 3.00<br>2.00<br>4.00 | indicating that the cubic<br>measurement is multiplied by 3 | |
| 2/ | 3.00<br>2.00 | indicating that the square metre<br>measurement is multiplied by 2 | |

The timesing figure is kept in the first column and separated from the dimension by a diagonal stroke. An item timesed can be timesed again, with each multiplier multiplying everything to the right of it, as follows:

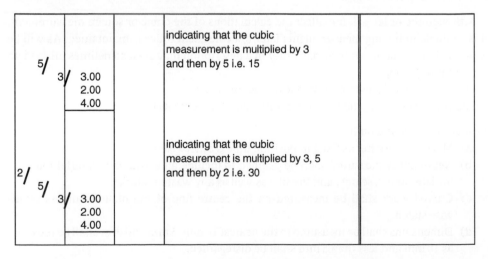

| | | | |
|---|---|---|---|
| 5/<br>  3/ | 3.00<br>2.00<br>4.00 | indicating that the cubic<br>measurement is multiplied by 3<br>and then by 5 i.e. 15 | |
| 2/<br>  5/<br>    3/ | 3.00<br>2.00<br>4.00 | indicating that the cubic<br>measurement is multiplied by 3, 5<br>and then by 2 i.e. 30 | |

The timesing is done to a linear or enumerated item in just the same way as shown here.

## Dotting on

In repeating a dimension, the taker-off may find that it cannot be multiplied but can be added. For instance, given that three items have been measured as follows:

| | |
|---|---|
| 3/ | 3.00 |
| | 2.00 |
| | 4.00 |

To make the train of thought clearer, what is called dotting on can be used as follows:

| | | indicating that the cubic |
|---|---|---|
| | | measurement is multiplied by 3 |
| | | plus 2 i.e. 5 |
| 2 · 3/ | 3.00 | |
| | 2.00 | |
| | 4.00 | |

The dot is placed below the top figure to avoid any possible confusion with decimals, although these are usually avoided in timesing. Figures dotted on should be lower than the last, just as each one timesed is usually higher; this makes more space available than if they were all written in a horizontal line.

Timesing and dotting on can be combined, as follows:

| | | indicating that the cubic |
|---|---|---|
| 2/ | | measurement is multiplied by 3 |
| | | plus 2 and then 2 i.e. 10 |
| 2 · 3/ | 3.00 | |
| | 2.00 | |
| | 4.00 | |
| | | indicating that the cubic |
| | | measurement is multiplied by 3 |
| | | and then 2 plus 3 i.e. 15 |
| 3 · 2 / 3/ | 3.00 | |
| | 2.00 | |
| | 4.00 | |

Care must be taken when writing fractions in the timesing column that the line dividing the numerator from the denominator is horizontal, thereby avoiding any confusion with timesing. It should also be noted that this technique is more useful where measurements are being recorded manually, i.e. recording dimensions on site and that if a spreadsheet or similar is being used, it is easier to overtype the multiplier.

## Waste calculations

Except in very simple cases, dimensions should not be calculated mentally. Not only will the risk of error be reduced if the calculations are written down, because they will be checked, but also another person can readily see the origin of the dimension. These preliminary calculations, known as *waste calculations*, are made on the right-hand side of the description column. They must be written definitely and clearly, and not scribbled as if they were a calculation worked out on scrap paper. The term *waste* used for this part of the column might be thought to imply 'useless', but in fact it implies 'a means to an end'. Every effort must be made to commit to writing the train of thought of the taker-off. Waste calculations should be limited to those necessary for the clear setting down of the dimensions by the taker-off and should not take the place of squaring.

An example of waste calculations is given in this chapter. These should not be written directly on the drawings. It is, however, necessary for dimensions to be traced from the drawings through any adjustments in the waste calculations to the figures used in the dimension column. They should preferably precede the description. An example of this is included along with the description for excavating a trench at the end of the 'Descriptions' section.

## Alterations in dimensions

Where a dimension has been set down incorrectly and is to be altered, either it should be neatly crossed out and the new dimension written in, or the word *nil* should be written against it in the squaring column to indicate that it is cancelled. Where there are a number of measurements in the dimension column, care must be taken to indicate clearly how far the nil applies. This may be done as follows:

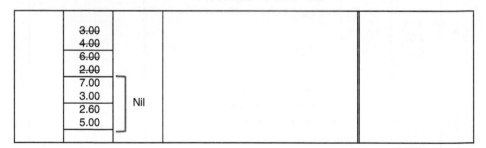

No attempt should be made to alter figures (e.g. a 2 into a 3, or a 3 into an 8). The figure may appear to have been altered satisfactorily, but it may look quite different to another person or when the dimensions are photocopied. Every figure must be absolutely clear, a page a little untidy but with unmistakable figures being far preferable to one where the figures give rise to uncertainties and consequent error. Deletions with correcting fluid should never be made as it is often of value to know what was written in the first instance. It is best to nil entirely and write out again any dimensions that are getting too confused by alterations; great care is needed in copying dimensions, or mistakes may be made, it being particularly easy to miss copying the timesing. Therefore, where any dimensions are rewritten, they should be checked very carefully against the original.

## The descriptions

The description of the item measured is written in the description column, on a level with its associated dimensions as follows (the waste calculations are also shown):

|  |  |  |  |  |
|---|---|---|---|---|
|  |  |  |  | <u>centre line</u> |
|  |  |  |  | 34.000 |
|  |  |  |  | 16.000 |
|  |  |  | 2 $\bigg/$ | 50.000 |
|  |  |  |  | 100.000 |
|  |  |  | <u>ddt</u> corner adjustment |  |
|  |  |  | 4/2/0.5/0.215 | 0.860 |
|  |  |  |  | 99.140 |
|  |  |  |  | <u>av. depth</u> |
|  |  |  |  | 15.050 |
|  |  |  |  | 12.000 |
|  |  |  |  | 13.500 |
|  |  |  |  | 11.700 |
|  |  |  |  | 52.250 |
|  |  |  | divide by | 4 |
|  |  |  |  | 13.063 |
|  |  |  | <u>ddt</u> |  |
|  |  |  | underside of conc | 12.050 |
|  |  |  |  | 1.013 |
|  |  |  | Excavation, starting topsoil reduced level, foundation excavation, not exceeding 2.00 m deep | |
| 99.14 |  |  | 5.6.2.1.0 | |
| 1.00 |  |  |  | |
| 1.01 |  |  |  | |

During the process of manual taking-off, descriptions are often written in shorthand. The contents of the description are normally established with reference to the standard method of measurement. An extract from the concrete section of NRM2 has been reproduced in Figure 1.

| Item or work to be measured | Unit | Level one | Level two | Level three | Notes |
|---|---|---|---|---|---|
| Plain in-situ concrete | | | | | |
| Reinforced in-situ concrete | | | | | |
| Fibre reinforced in-situ concrete | | | | | |
| Sprayed in-situ concrete | | | | | |
| 1 Mass concrete. | m³ | 1 Any thickness. | 1 In filling voids. <br> 2 In trench filling. <br> 3 In any other situation, details stated. | 1 Poured on or against earth or unblinded hardcore. | 1 Mass concrete is any unreinforced bulk concrete not measured elsewhere. <br><br> 2 The volumes of each type of mass concrete work may be aggregated or given separately. |
| 2 Horizontal work. | m³ | 1 ≤ 300mm thick. <br> 2 > 300mm thick. | 1 In blinding. <br> 2 In structures. | 1 Poured on or against earth or unblinded hardcore. <br><br> 2 Reinforced > 5%. | 1 Horizontal work includes blinding, beds, foundations, pile caps, column bases, ground beams, slabs, coffered and troughed slabs, landings, beams, attached beams, beam casings, shear heads, upstands whose height is less than or equal to three times their width, kerbs and copings. <br><br> 2 The volumes of each type of horizontal work may be aggregated or given separately. <br><br> 3 Work laid in bays should be so described giving average area of bays. |

**Fig. 1**  *Source*: From Section 11 of NRM2, by kind permission of the RICS.

A heading would be required identifying the kind and quality of the concrete; any tests of materials and finished work; measures to achieve water-tightness; limitations on the method, sequence, speed, or size of pouring; and any methods of compaction and curing that might be specified. This can be found from the additional specification rules at the start of the concrete Section 11.

Typically, a description for a specific item will be built up as follows:

- Location of the work: from the first column.
- The second column identifies the unit of measurement to be used.
- Depending on the item to be measured, further dimensions or descriptions required from columns headed Level 1, Level 2, and Level 3.
- Any further information can be taken from the fifth column, where applicable.

A typical description of concrete in a suspended floor slab would be:

| | | | | |
|---|---|---|---|---|
| | | | | Teaching note. |
| | 4.00 <br> 9.00 <br> 0.20 | | Reinforced insitu concrete, horizontal work, ≤ 300 mm thick, in structures <br> 11.2.1.2.0 | measured in m³ so 3 dimensions in dimension column |

If two or more measurements are related to one description, it is normal practice for them to be bracketed together as follows:

| | | |
|---|---|---|
| 99.14<br>1.00<br>0.40<br><br>3.05<br>0.76<br>0.45 | Reinforced insitu concrete,<br>horizontal work, > 300 mm thick,<br>in structures, poured on or<br>against earth or unblinded<br>hardcore<br>11.2.2.2.1 | |

The bracket is placed on the outside of the squaring column. A clear indication is made of where the bracket ends, the bracket itself usually being a vertical line with a short cross mark to indicate top and bottom. A full description of the measurement rules can be found in NRM2 under section 3.2.

## Anding-on

Where two or more descriptions are to be applied to one measurement, they are written as follows:

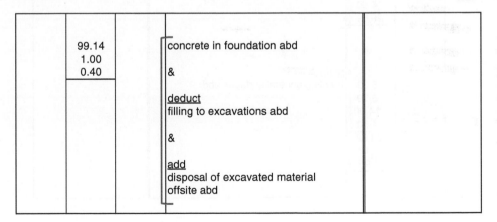

| | | |
|---|---|---|
| 99.14<br>1.00<br>0.40 | concrete in foundation abd<br><br>&<br><br><u>deduct</u><br>filling to excavations abd<br><br>&<br><br><u>add</u><br>disposal of excavated material<br>offsite abd | |

In this case, each description is separated by an ampersand '&' on a line by itself. Care must be taken when 'anding-on' in this way so as not to confuse a square metre item with a linear item. It sometimes happens when manually taking-off that it is very convenient to do so, but the distinction must be made quite clear so that the linear quantity is not used instead of the square metre. For example:

| | | | | |
|---|---|---|---|---|
| | 5.33 | 5.33 | Skirtings, 25 × 100 mm | |
| | 5.79 | 5.79 | 22.1.1.1.0.0 | |
| 2/ | 7.00 | 14.00 | | |
| | 4.50 | 4.50 | & | |
| | | 29.62 | | |
| | | | deduct | |
| | | | Painting general surfaces abd | |
| | | | × 0.100 = 2.96 m$^2$ | |

In this example, if a spreadsheet is being used, the dimensions could be cut and pasted below and the height added as shown. This has the benefit of showing the total in the squaring column where it would usually occur.

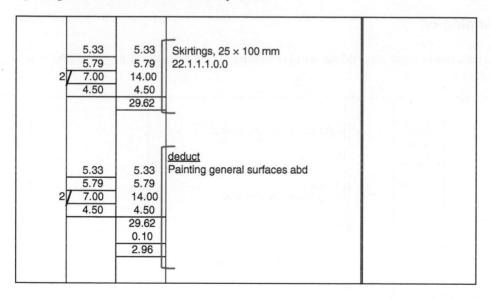

| | | | | |
|---|---|---|---|---|
| | 5.33 | 5.33 | Skirtings, 25 × 100 mm | |
| | 5.79 | 5.79 | 22.1.1.1.0.0 | |
| 2/ | 7.00 | 14.00 | | |
| | 4.50 | 4.50 | | |
| | | 29.62 | | |
| | | | | |
| | | | deduct | |
| | 5.33 | 5.33 | Painting general surfaces abd | |
| | 5.79 | 5.79 | | |
| 2/ | 7.00 | 14.00 | | |
| | 4.50 | 4.50 | | |
| | | 29.62 | | |
| | | 0.10 | | |
| | | 2.96 | | |

To measure the deduction of the emulsion paint in this way saves setting down all the dimensions again as square metre items (exactly the same lengths being used). Similarly, a square metre item might be marked to be multiplied by a third dimension to make a cube, for example:

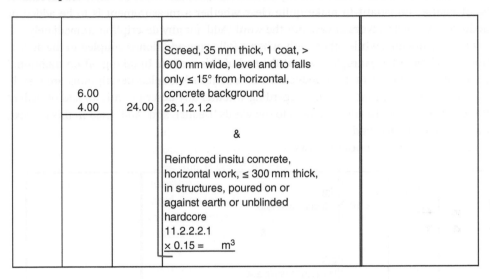

| | | | |
|---|---|---|---|
| 6.00<br>4.00 | 24.00 | Screed, 35 mm thick, 1 coat, ><br>600 mm wide, level and to falls<br>only ≤ 15° from horizontal,<br>concrete background<br>28.1.2.1.2<br><br>&<br><br>Reinforced insitu concrete,<br>horizontal work, ≤ 300 mm thick,<br>in structures, poured on or<br>against earth or unblinded<br>hardcore<br>11.2.2.2.1<br>× 0.15 =     m³ | |

This example can also be readily adapted for use with a spreadsheet by copying the dimensions and pasting against the description as below:

| | | | |
|---|---|---|---|
| 6.00<br>4.00 | 24.00 | Screed, 35 mm thick, 1 coat, ><br>600 mm wide, level and to falls<br>only ≤ 15° from horizontal,<br>concrete background<br>28.1.2.1.2 | |
| 6.00<br>4.00<br>0.15 | 3.60 | Reinforced insitu concrete,<br>horizontal work, ≤ 300 mm thick,<br>in structures, poured on or<br>against earth or unblinded<br>hardcore<br>11.2.2.2.1 | |

This approach is particularly useful for the manual measurement of ground floor slabs, for example where you may have a large number of complex dimensions all relating to the various items associated with surfacing.

## Cross-references

The taker-off should try to ensure that dimensions are clear to others, as it is quite possible that, when variations on the contract have to be adjusted, someone else will be entrusted with the work and will have to find their way around the dimensions. Cross-referencing between the drawings and the dimensions is essential for ease and speed in locating why particular dimensions have been used. It has been known for projects to be postponed for a year or two after tenders were received, and even the taker-off will then need some references and notes to refresh the memory. The dimensions may also need to be referred to when preparing the final account, and again clarity is important.

## Clearness of the dimensions

Besides the use of cross-references, a good deal can be done to make the dimensions clear by the manner in which they are set down. It has already been pointed out that a regular order of length, width, and depth (or height) should be maintained in writing down the dimensions; even when it may be difficult to determine which is length or width, a consistent order should be kept. In measuring areas of floor finishes, for instance, the dimensions horizontal on the plan could be put first, followed by the vertical ones. Calculations should be made as waste on the dimension paper and not mentally, and timesing should be done consistently.

For instance, in measuring six doors each with four squares of glass, all timesed for two floors, the dimensions should be timesed as follows:

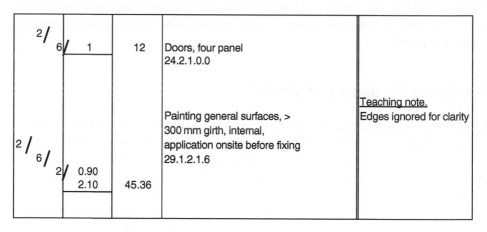

with the last item not being written as follows:

| | | | |
|---|---|---|---|
| 6/ 2/ 2/ | 0.90 2.10 | 45.36 | Painting general surfaces, > 300 mm girth, internal, application onsite before fixing 29.1.2.1.6 | |

When timesing becomes complicated, it will help considerably in tracing items if the method of timesing is consistent. In the example given here, the outer timesing represents the floors, and the next the number of doors; it could be confusing if the order is reversed in the middle of a series. The use of coloured pens in timesing or dotting on to represent different floors or sections of the work will be found to help considerably in tracing dimensions later.

## Headings

The use of headings in the dimensions will further aid future reference. Apart from the taking-off section heading already suggested for each page, subheadings or *signposts* should be clearly written wherever possible, and these will stand out if underlined. The sequence of measurement in the section will then easily be followed by a glance at the subheadings.

## Notes

The making of notes by the taker-off on the dimensions is of the utmost value. Such notes are usually for reference before the taking-off is finished (e.g. notes of items to measure or queries to be settled).

This type of note should be written at the right-hand side of the description column, and it is best separated by a line or bracket to prevent confusion with descriptions. There is often a column provided on data entry sheets for notes or comments to be added.

The type of note, made for reference before the taking-off is finished, is necessary when perhaps some point must be referred to the architect or engineer, or for other reasons something cannot be finally measured. As mentioned, a list of such queries should be compiled. *To take* notes are entered in dimensions when a taker-off dealing with one section feels that an item, although arguably within the work being measured, is better taken with another section (e.g. tile splashbacks to sinks taken with the plumbing services rather than with the finishes or vice versa). These *to take* notes should be written clearly in the dimensions and collected together before the bill is prepared, and a check should be made to ensure that all items listed have been measured. Memoranda, too, should be made at the end of the day's work of anything unfinished which, the train of thought being broken, might be forgotten. It is much safer to make notes of such matters on the dimensions than to trust them to memory, and these notes enable the work to be carried on in the case of unexpected absence. Such notes as these are sometimes written in pencil to be erased when dealt with, or they may be written in ink across the description column in such a way that they cannot be missed.

## Insertion of items

It quite often happens that the taker-off must go back to the work and make alterations or insert additional items, either because items have been forgotten or because revised details have been received, as described in this chapter. Such alterations and additions should

# 5
# Alternative Systems

This book concentrates on the traditional taking-off in order to explain the basic principles which can then be applied to alternative measurement and billing systems.

This chapter briefly outlines the alternative systems that have been developed and adopted in order to achieve standardisation, increase the efficiency of bill production, and eliminate time-consuming data-processing activities.

Computerisation of the measurements and bill preparation processes have been evolving over a period of time based on the use of standardised descriptions and advances in computer hardware and software.

## Standardisation

The art of taking off is to a large extent based on developing a systematic approach, having a sound knowledge of building construction, and acquiring the mathematical skill to calculate and measure dimensions and the ability to write clear descriptions. Of these, probably the latter is the most difficult to master because it has to be built up from long experience, especially in estimating and dealing with errors and claims arising from inadequate descriptions. The taker-off has to ensure that descriptions comply with the requirements of the Standard Method of Measurement (SMM) and refer properly to British Standards, and often has to frame them under pressure of time from complex and sometimes incomplete drawings and specifications. It is no wonder, therefore, that occasionally errors and omissions occur in descriptions, and it is not unknown for there to be diversity between different bills from the same office or even between different sections within the same bill, particularly in elemental formats. Furthermore, both personality and experience influence the taker-off's approach to compiling descriptions; and brevity, verbosity, and English literacy all have their effect.

In former years, an assistant became proficient in framing descriptions by experience gained in a working-up section preparing abstracts and bills. With the introduction of computerised systems coupled with longer full-time education, this form of experience has been lost and alternatives have had to be established.

*Willis's Elements of Quantity Surveying,* Fourteenth Edition. Roy Hills and Sandra Lee.
© 2024 John Wiley & Sons Ltd. Published 2024 by John Wiley & Sons Ltd.

Standardisation was achieved either by compiling a standard bill or library within an individual organisation, including all the descriptions likely to be encountered in the type of work usually dealt with, or by using one of the published standard libraries.

## Standard libraries

Most libraries rely on the fact that phrases within descriptions are frequently repeated, whereas the complete description is often peculiar to one situation. Therefore, through splitting descriptions into phrases and allocating the phrases to various levels, full descriptions can be built up by selecting phrases at each level, some being essential and others optional. Considerable saving in bulk is achieved as phrases that are frequently used are listed once only.

The levels of phrase in a standard library could be as follows:

- Level 1: main work section.
- Level 2: subsidiary classification.
- Level 3: main specification in heading.
- Level 4: description of item.
- Level 5: size and number.
- Levels 6 and 7: written short items (extra over items).

By selecting obligatory and optional headings and phrases from each level, combinations are built up to complete the description. The advantages of such a system are numerous; some of them are summarised in the following list:

- Descriptions are standardised and consistent.
- It provides an *aide-mémoire* for the taker-off.
- Billing can be carried out by less trained staff.
- Bill editing is greatly reduced.
- Consistency between bills aids price comparison of items.
- The simplest possible wording, almost in note form, avoids ambiguities.
- The use of 'ditto' (often causing doubt) is avoided.
- It complies with the SMM.
- Consistency between bills from various sources aids the estimator.

This list would appear to give standard libraries an overwhelming advantage. There is, however, a danger that the taker-off will apply a readily available standard description to the item being measured even though it does not fit exactly, rather than compiling carefully a proper description; such an item is commonly called a *rogue item*. Furthermore, there is no doubt that if the taker-off has to keep referring to bulky documents, the train of thought is interrupted, and speed is reduced.

## Computerised bill production

With rapid advances in computer technology, and as computer hardware and software applications have become comparatively less expensive, many firms have adopted one of the many computer systems now available. They have become widely used not only in bill production but also for assistance in carrying out most other quantity surveying functions. It is desirable that the operation of these other functions be carried out without having to re-enter the data stored for bill production, and therefore database systems are used.

A wide variety of bill of quantities production systems are now available, which have varying degrees of sophistication.

Although paper-based taking-off can be manually entered into the systems, the majority now available have the facility to input dimensions directly by using either the keyboard or a digitiser. A digitiser is an electronically sensitive drawing board from which dimensions may be electronically scaled from the drawings into the system. Lengths, perimeters, and areas of both simple and complex shapes are calculated automatically. All systems are based on having a standard library of descriptions built up from library text displayed on the screen. The problem facing the industry is that there are a number of different standard libraries used by each supplier of bill preparation software, and often each company or practice that uses the software creates its own library from scratch. Some companies use the library provided with the software as a starting point and then add their own rogue items that they frequently use. When a quantity surveyor moves between companies, he or she will find the basic principles of the various systems similar, but there will be a need for re-training on the different software packages to ensure that the full features of the system are understood and utilised.

When a particular description is not available from the standard library, there is a facility to create rogue items. The number of rogues is likely to be more in a complex building project than in a simple one. It is therefore the ease with which creating and sorting rogues can be carried out that determines how efficient a particular system is.

An extract from a typical library of descriptions is shown in Figure 3. This figure shows a sample of the various levels of the description that are available, along with a coding system. The main heading might only be used once in a bill of quantities, but the common brickwork may well be required in both 102.5 and 215 mm thicknesses. If you look at the section in NRM2 for brickwork and blockwork, you will see an extensive list of potential items that would need to be included in a library database.

| Code | Description | Unit |
|---|---|---|
| F10 | F10 BRICK / BLOCK WALLING | |
| | | |
| F1001 | COMMON BRICKWORK | |
| F1003 | FACING BRICKWORK | |
| | | |
| F10 - - 001 | Walls | |
| F10 - - 003 | Isolated piers | |
| | | |
| F10 - - - - - 01 | 102.5 mm thick | m$^2$ |
| F10 - - - - - 02 | 215 mm thick | m$^2$ |
| F10 - - - - - 03 | .... mm thick | m$^2$ |
| | | |
| F10 - - - - - - - 01 | battering | |
| F10 - - - - - - - 02 | tapering one side | |
| F10 - - - - - - - 03 | tapering both sides | |

**Fig. 3**

Correctly coding items measured on paper for subsequent entry into a measurement package is of paramount importance, and there is always the danger of mis-coding. A record is kept, however, and can be physically checked. The code below (F10) is from the Common Arrangement of Work Sections (CAWS) which was used in SMM7 but is not used in NRM2. The Construction Project Information Committee (CPIc) continues to develop and promote standard coding for production information in the construction industry.

In contrast, some packages allow you to search through the library of descriptions to choose the item to include in your take-off, and it is less likely an error will be made at this stage. Dimensions can then be entered with the relevant description along with any waste calculations or side notes. A continual print of the data entry is useful so that you have a record of the logic used for the taking-off. This can prove very useful at the post-contract stage when researching how the bill was measured so that variations can be assessed.

The order of taking-off for direct computer entry may differ from that used in the traditional approach. Utilising the 'anding-on' facility, or as some suppliers of software call it 'carry forward', can be extremely useful to save re-entering a long stream of dimensions. Before you start a section of measurement, think carefully about how you want to order your work.

Common features with software packages are the facility to utilise data from previous bills of quantities, to calculate the weight of steel and reinforcement automatically, and to display rules for calculating volumes. Systems can contain standard libraries for different standard methods of measurement and, being database systems, they have a multiple sort facility and can produce different bill formats as required. They can also sort, display, and print data to different levels of detail. Draft bills can be produced for editing, and in many ways the editing process is more important than when using manual methods of bill production.

The development of bill production systems is ongoing, and these systems are continually being upgraded. Many systems already have the facility to generate bill items directly from computer-aided design (CAD) data or other schedule data (i.e. pdf) by combining a drawing reader function and spreadsheet which allows dimensions to be transferred from a drawing to a BQ or estimating template. As such fully computerised systems continue to be developed and become more widespread, the quantity surveyor will concentrate more on interpreting the data produced, ensuring their completeness and utilising them not only for tender documentation production but also for cost control and post-contract administration purposes.

Electronic data interchange (EDI) is also becoming more widely used: the bill is issued in electronic form to the contractors, who can then use it to price their tender for the work. The advantages of this are that it speeds up the transmission and receipt of information, reduces the paperwork involved, and eliminates the wasteful repetition of rekeying information into a separate system (which is time consuming and error prone).

## E-tendering

The process of e-tendering is where the tender for a project is managed entirely electronically over the internet.

The Royal Institution of Chartered Surveyors (RICS) has published a guidance note for their members on what a system should involve and how it should be managed. This puts forward the proposition that the primary strategic purpose behind e-tendering is the desire to drive down costs. There are, however, other benefits that would support its wider use, such as the potential for aggregation of buying power, the elimination of waste, the simplification of the process, and a fairer assessment between tenders.

Without the need for the production of multiple copies of paper-based information, there is a less environmentally demanding and more sustainable tender process.

The impact of e-tendering on measurement is that bills of quantities need to be produced electronically, and it should be possible for them to be converted into a commonly accepted format by contractors for pricing.

The RICS sees the barriers to successful implementation of e-tendering as being a lack of agreed standards, little or no impartial advice, and the potential legal and technical traps. Hence, there is a very slow move towards its acceptance.

Space is not available here to discuss the process in detail, and those wishing to obtain further information can do so through the RICS website.

## Site dimension books

When surveyors are visiting a site, it helps to have a notebook that is already ruled up for measurement. Typically, these notebooks are just slightly smaller than A5, so that they are easily held, and have a hard cover to protect them throughout the life of a project.

The pages are ruled so that dimensions can be set down in the traditional format as shown:

| | |
|---|---|
| Date: | |
| Project: | |
| Notes on purpose of measure, etc. | |
| | Descriptions and waste calculations |

As site visits may not occur every day, it is important to note the date that a measurement was made and the purpose of the measurement. It may be to record details of a variation, or a measure of work for interim payment. Sketches, drawing references, and location notes within the building to identify precisely where the work is located should also be made.

The site dimension book should be numbered for the project, as there may well be quite a few books over the life of the project, and the name of the surveyor recorded at the front. These books should then be filed when the project is complete so that there is a record should there be a dispute during agreement of the final account.

## Estimating paper

Slightly different paper again is used when an approximate estimate is being prepared so that the dimensions and costing can occur on the same page. This type of paper can be purchased either loose by the ream or in A4 pads. A typical page would be follows:

| 1 | 2 | 3 | | 4 | 5 | 6 | Currency |
|---|---|---|---|---|---|---|---|
| | | | Waste calculations | | | | |
| | | | | | | | |
| | | | Description | | | | |
| | | | | | | | |
| | | | Cost build up | | | | |
| | | | | | | | |
| | | | | | | | |

Columns 1–3 are the traditional dimension columns, column 4 is the total quantity measured, column 5 is the unit of measure, and column 6 is the rate to be used to price the item. Spreadsheet software could also be used to prepare a standard template.

# 6

# Preliminary Calculations

## Mathematical knowledge

It is assumed that the reader is acquainted with geometry and trigonometry, knowledge of which, as of building construction, is an essential prerequisite to the study of quantity surveying. It is rare that more detailed knowledge is required than the properties of the rectangle, triangle, and circle. Less-known formulae are detailed in the Appendix and can be looked up in an appropriate mathematical reference book. The properties of the rectangle, triangle, and circle should, however, be thoroughly understood; if a student is not acquainted with them and with elementary trigonometry, it is recommended to study these before going any further. This chapter shows some examples of how the theoretical knowledge of mensuration is applied to building work. Wherever possible, lengths should be found by calculation from figured dimensions on the drawings, rather than by scaling. Where scaling has to be used, a check should be made to ensure that other figured dimensions are accurate, as some reproduction methods affect the scale. If using computer-aided design (CAD) drawings, a check on the scale should be made and the drawings calibrated.

## Perimeter of buildings

Figured dimensions may be given internally or externally, and either can be worked from. Figure 4 shows a plain rectangular building with one brick wall 215 mm thick.

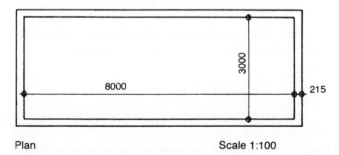

Plan                                        Scale 1:100

**Fig. 4**

*Willis's Elements of Quantity Surveying,* Fourteenth Edition. Roy Hills and Sandra Lee.
© 2024 John Wiley & Sons Ltd. Published 2024 by John Wiley & Sons Ltd.

Here, internal dimensions have been provided and are used to calculate the girth of the wall. This is presented as a waste calculation with the length first and then the width. These are added together and then multiplied by two to obtain the internal girth of the wall.

| | | | | |
|---|---|---|---|---|
| | | | 8.000 | |
| | | | 3.000 | |
| | | 2/ | 11.000 | |
| | | | 22.000 | |

The length of the perimeter of the external face of the walls, which may be required for the measurement of external rendering, can be calculated by adding twice the thickness of the wall at each corner.

| | | | | |
|---|---|---|---|---|
| | | ab | 22.000 | |
| | | 4/2/0.215 | 1.720 | |
| | | | 23.720 | |

This can be confirmed by adding the thickness of the wall to each internal dimension first.

| | | | | |
|---|---|---|---|---|
| | | | width | |
| | | | 3.000 | |
| | | 2/0.215 | 0.430 | |
| | | | 3.430 | |
| | | | | |
| | | | length | |
| | | | 8.000 | |
| | | 2/0.215 | 0.430 | |
| | | | 8.430 | |
| | | | | |
| | | | 3.430 | |
| | | | 8.430 | |
| | | 2/ | 11.860 | |
| | | | 23.720 | |

It will be seen that twice the thickness of the wall has been added for each corner. This approach is the basis of the calculation that can be used to find the centre line of the wall, which is explained in the next section.

## Centre line of the wall

NRM2 requires a brick or block wall to be measured net in square metres along its centre line, stating the thickness of the wall. The measurement of items such as foundation trenches, concrete footings, and forming cavities all use the same centre line for the length when calculating their quantities. As it is the net measurement that is required, there should be no duplication at the corners of a building. By looking again at Figure 4, it can be seen that two walls of 8.43 m and two walls of 3.00 m are required, giving a net length around the centre line of 22.86 m. This should be apparent just by looking at the plan; however, most plans are not a simple rectangular shape, and waste calculations will be necessary to establish the required centre line. The detail of a corner is shown in Figure 5, showing that the centre line requires either the thickness of the wall to be added or deducted from the internal or external dimensions, respectively.

A straightforward way of calculating the centre line measurement or mean girth from the internal dimensions is as follows:

| | | | | |
|---|---|---|---|---|
| | | | internal girth ab  22.000 | |
| | | | add corner adjustment | |
| | | | 4/2/0.5/0.215    0.860 | |
| | | | centre line of wall  22.860 | |

It will be seen that twice of half the wall thickness is added to the internal perimeter for each external corner, that is, 4 × 2 × the distance moved (half the thickness of the wall in this instance). If you remember this formula, it can be applied in numerous situations during measurement.

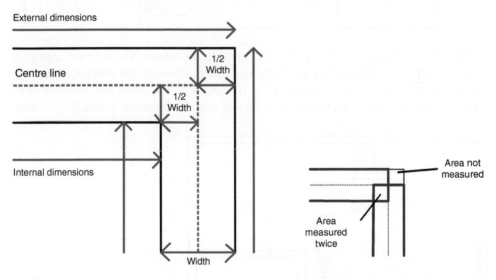

**Fig. 5**

The whole process can, of course, be reversed. If the external perimeter is taken instead of the internal, by deducting instead of adding, the mean length of the wall is obtained as follows:

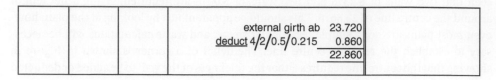

|  |  |  | external girth ab | 23.720 |  |
|---|---|---|---|---|---|
|  |  |  | deduct 4/2/0.5/0.215 | 0.860 |  |
|  |  |  |  | 22.860 |  |

By becoming familiar with the centre line calculation, more complex-shaped buildings can be measured quickly, and repeated use made of the same dimension. It is therefore important that this dimension is correct as a simple error can be repeated through many items in a bill of quantities. If the shape of the building is slightly more complicated, as shown in Figure 6, the same principle may be applied to the calculations. For example, the centre line or mean girth measurement is:

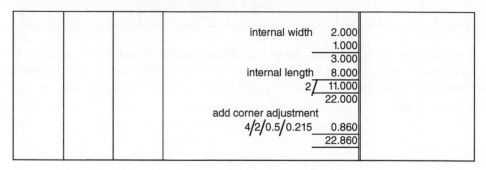

|  |  |  | internal width | 2.000 |  |
|---|---|---|---|---|---|
|  |  |  |  | 1.000 |  |
|  |  |  |  | 3.000 |  |
|  |  |  | internal length | 8.000 |  |
|  |  |  | 2/ | 11.000 |  |
|  |  |  |  | 22.000 |  |
|  |  |  | add corner adjustment |  |  |
|  |  |  | 4/2/0.5/0.215 | 0.860 |  |
|  |  |  |  | 22.860 |  |

Where the wall breaks back (enlarged in Figure 7), the internal and external angles balance each other. As before, if the inside face has been measured, the external angle needs the thickness of the wall to be added to give the length on the centre line, whereas in the case of the internal angle. the thickness of the wall must be deducted. In fact, the perimeter

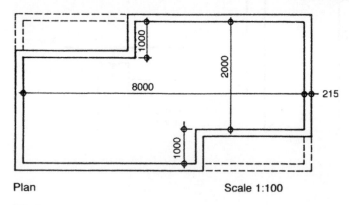

Plan                                Scale 1:100

**Fig. 6**

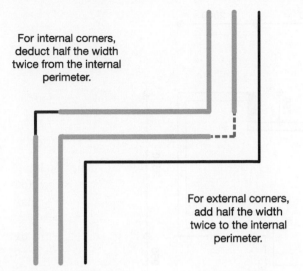

For internal corners, deduct half the width twice from the internal perimeter.

For external corners, add half the width twice to the internal perimeter.

**Fig. 7**

of the building is the same as if the corners were as shown dotted in Figure 6. In short, to arrive at the mean girth, the thickness of the wall must be added for every external angle in excess of the number of internal angles, i.e. external angles minus internal angles equals multiplication factor.

This collection of the perimeter of walls being of great importance, one further and more complicated example is given in Figure 8. This time, the calculations are made from the external figured dimensions instead of from the internal ones.

When smaller dimensions are given, you could check that their total equals the overall dimensions given, for example in relation to Figure 8 this would be:

| <u>top</u> | <u>bottom</u> |
|---|---|
| 2.000 | 3.600 |
| 2.750 | 0.900 |
| 2.000 | 2.250 |
| 6.750 | 6.750 |

| | <u>right</u> |
|---|---|
| | 0.950 |
| | 0.950 |
| | 3.650 |
| | 5.550 |

| | 1.200 | |
|---|---|---|
| deduct | 0.750 | 0.450 |
| | | 6.000 |

| | <u>left</u> | |
|---|---|---|
| | | 6.000 |

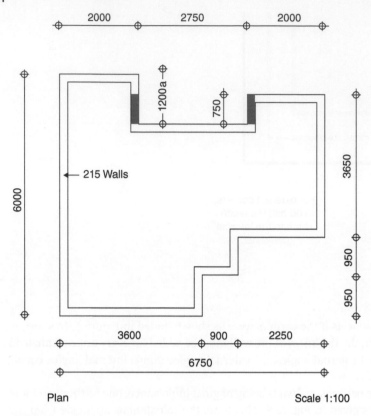

Plan                                         Scale 1:100

**Fig. 8**

Then, the calculation of the centre line is as follows:

| | | | | | |
|---|---|---|---|---|---|
| | | | external length | 6.750 | |
| | | | external width | 6.000 | |
| | | | 2⟋ | 12.750 | |
| | | | | 25.500 | |
| | | | add recess sides | | |
| | | | 2/0.750 | 1.500 | |
| | | | | 27.000 | |
| | | | deduct corner adjustment | | |
| | | | 4/2/0.5/0.215 | 0.860 | |
| | | | | 26.140 | |

If you have difficulty in appreciating why 2/0.750 is added, then colour the lengths of the wall that have already been measured carefully. This can make the extra length obvious. You could also think of this shape as having flexible joints, and 'push' the shape out to form a rectangle. Again, you will be left with two additional lengths that will not 'fit', these being the 2/0.750 for the 'indent' or 'recess' marked 'a' on Figure 8.

The accurate calculation of the centre line is a most important one, as, having once been made, it is often used not only for several items in foundation measurement but also for brickwork and facings, with possibly copings, string courses, and so on.

When calculating the girths of the individual components of a cavity wall, the same process should be followed. First, the girth of the perimeter can be calculated; then, progressively, the mean girths of the outer skin, the cavity, and then, finally, the inner skin (as in Example 3 in Chapter 9).

# 7

# General Principles for Taking-Off

In this chapter, suggestions are given regarding procedure, general principles of measurement, and other information that is necessary for the guidance of those responsible for coordinating the taking-off process, for the individual takers-off, and for the study of the examples that follow.

It must not be forgotten that the methods of different individuals and the customs of different practices vary in the organisation and execution of this work. The fact that a particular suggestion is made here does not mean that it is universally adopted, nor does it preclude the use of an alternative.

A systematic approach is essential to ensure that the precise documents used when preparing a bill are known along with any revision numbers and the date of receipt. Recording who is dealing with which elements and what copies of drawings they are using helps with the management of the process, as various members of staff will refer to this record during the period of the project and not just during the bill preparation stage. When variations are required at the post-contract stage or if a contractor challenges a quantity, the dimensions will need to be referred to.

## Receipt of the drawings

If physical drawings are received, they should be stamped with the date and entered on a list showing the drawing number, title, date of preparation, date received, and number of copies. A drawing register will often accompany any drawings issued and this should also be checked against the actual drawings received and any discrepancies queried. This list forms a useful reference and shows when any revisions to previous drawings were received. A check should be made with the architect and engineer to ascertain what further drawings are under preparation that are likely to be received before the bill of quantities is completed. At the completion of the taking-off, each drawing used should be stamped 'used for quantities', which avoids confusion later on as to whether or not subsequent revisions were incorporated in the bill. When using computer-aided design (CAD) drawings, the version of the drawing used is just as important, and a schedule of drawings used for the bill is still required.

*Willis's Elements of Quantity Surveying,* Fourteenth Edition. Roy Hills and Sandra Lee.
© 2024 John Wiley & Sons Ltd. Published 2024 by John Wiley & Sons Ltd.

## Preliminary study of drawings

Before any dimensions are entered or written at all, the taker-off should look over the drawings and study the general character of the building.

A check should be made to ensure that all floor plans and elevations are shown and that the positions of the sections are marked on all the plans. A more detailed inspection should be made to ensure that the windows, doors, rainwater pipes, etc., shown on the plans are also shown on the elevations and vice versa. When drawings have been prepared by engineers, they should be compared with those prepared by the architect to ensure that there are no discrepancies in the layout or dimensions. If overall dimensions are shown, these should be checked against the total of room and wall measurements. If overall dimensions are missing, these should be calculated and marked on the plans, and all projections on external walls should be dimensioned. If this is done, the calculation of the mean girth and of the perimeter of the external walls is simplified. If hard copies of the drawings are being used, it can be an advantage to dimension every room on the plans except where a series of rooms are obviously all of the same dimension in one direction. A little time spent in this preliminary figuring of the drawings obviates the possibility of inconsistency in dimensions. Where more than one taker-off is employed, the result of this work must be communicated to each of them, and their drawings marked accordingly. Figured dimensions must always be followed in preference to scaled, and any dimensions that can be calculated from those figured should be worked out. A larger scale drawing will usually override a smaller scale one, except when the smaller is figured and the larger is not. Practices vary when CAD drawings are used; however, enlarging drawings sufficiently to obtain an accurate measure is important.

## Queries with the designers

Preliminary inspection of the drawings may also give rise to a number of queries to be raised with the architect or engineer, relating to missing information or discrepancies. Settling questions at an early stage saves interruption to the taking-off and consequently increases productivity. It is important that all queries are set down and answered in writing by for example, using a Request For Information (RFI) as illustrated below. Queries should be listed on the left-hand side of a sheet of paper, with the answers, when received, shown opposite on the right-hand side, together with the date and source of the answer (Figure 9). The quantity surveyor, however, may make assumptions for any gaps in design information, and these should be confirmed by the appropriate designer before the BQ is finalised. These query lists are sent to the architect or the engineer for completion.

If the survey of the existing site does not show levels, or if those that are shown are insufficient, it is prudent to request that a grid of levels is taken over the site. These levels will prove invaluable when measuring earthworks and are essential for reference later if there is any question of remeasurement.

```
┌─────────────────────────────────────────────────────────────┐
│ Project Nr . . . . . . . . .    Southtown School    Sheet Nr . . . . . . . . . │
│                                    Queries                    │
│                                       │                       │
│ Date                                  │                       │
│                                       │                       │
│ Ref          Query                    │         Reply         │
│                                       │                  Date │
│ 1.  Finish to floor of entrance hall  │                       │
│     specified wood block,             │                       │
│     coloured as tile?                 │                       │
│                                       │                       │
│ 2.  Should dimension between          │                       │
│     piers on north wall be 5.08 not   │                       │
│     5.03? (to fit the overall 56.85)  │                       │
│                                       │                       │
│ 3.  Dpc not mentioned in the          │                       │
│     spec. notes?                      │                       │
│                                       │                       │
│ 4.  Should brick facing to            │                       │
│     concrete beams be tied back?      │                       │
│                       Etc. etc.       │                       │
└─────────────────────────────────────────────────────────────┘
```

**Fig. 9**

## Initial site visit

Before actually starting on the measurement, it is advisable to make an initial visit to the site. If nothing else, such a visit allows one to get the feel of the job and to be able to visualise later the various site references when they are encountered on the drawings. A further visit or visits will be necessary when the measuring has progressed in order to pick up the site clearance items (see Chapter 19). The initial visit will give an indication of what will be required in due course; in this respect, other items to be noted include such matters as boundaries generally (state and ownership of fences, walls, gates, etc.), the existence or otherwise of overhead and underground services, means of access, adjoining buildings (historic, civic, defence, etc.), and all matters that will be required for the subsequent drafting of the preliminaries bill (see Chapter 20). In addition, this visit is an opportunity to take the levels referred to here, and also to examine and note any trial holes that have been dug and are open for inspection. A photographic record of the site or specific parts of it may prove useful as a reminder of existing site conditions.

## Where to start

If all drawings are completed and available, the taker-off may follow the order of sections given in NRM2 or such other order as may be the custom of the office. This order will be seen to follow, more or less logically, the order of erection of the building, except that certain special work and services are dealt with at the end. It may, however, be that the design of, say, reinforced concrete foundations is not completed, and therefore measurement cannot begin as usual with substructure. In such a case, a point, such as damp-proof course level, must be selected from which to measure the structure; the work below that level will be measured at a later stage when the necessary information is available. It may be that 1:100 scale drawings have been received but 1:20 scale details are to follow. In such a case, internal finishes could probably be taken first, as the measurement of these is dependent generally upon the figured dimensions given on the 1:100 scale, which will probably not be altered. Moreover, these sections give the taker-off a good idea of the layout and general nature of the building. Specific forms of construction (e.g. a steel or concrete frame) may require a special sequence of measurement to be devised.

## Organising the work

Where several takers-off are involved in the measurement of work and a deadline has to be met for the completion of the bill, it is important that the team leader organises the work carefully. To enable progress to be monitored, a schedule should be prepared showing the sections of work to be measured, with the name or initials of the taker-off responsible, together with the target and actual dates of commencement and completion of the work. Clear instructions must be given to each member of the group, carefully defining the extent of the work to be taken in each section. Proper arrangements should be made for the collection of queries for the architect or engineer, and these should be edited by the team leader before being passed to the other consultants. Proper supervision should be made of junior staff involved, and due allowance made for staff leave and their commitment to other work.

## Taking-off by work sections

Some surveyors make a practice of taking-off by work sections instead of by elements of the building as described in this chapter. As the final bill is usually arranged in work sections, this system can minimise the sorting or abstract, which, when measuring by sections of the building, is necessary to collect and classify the items into work sections. Such a system may present difficulties in some cases, and it would seem to increase the risk of duplicating items or of forgetting something. When measuring by sections of the building, the taker-off mentally erects the building step by step and is less likely to miss items. Sometimes, a combination of the two systems may be used to advantage.

## Drawings

As mentioned in the 'Preliminary Study of Drawings' section, figured dimensions on drawings should be used in preference to scaling. Naturally, where there is a large discrepancy between the two, one should compare with other dimensions given, or, if this fails to produce a solution, consult the architect. Care must be taken to use the correct scale when taking measurements from drawings, particularly where a variety of scales has been used. It should also be noted that drawings often carry the instruction that dimensions should not be scaled. As items sometimes are measured from drawings, it is often beneficial to loop through the applicable written notes and perhaps colour in the work on the drawing. When using CAD drawings, most software will provide coloured drawings to depict the items measured. Print-offs of these are useful when checking or editing the final bills. An examination of the drawings marked up in this way, at the end of measurement, will soon reveal any items not measured. Not all measurable items are shown on drawings, particularly if the drawings are incomplete or in any case for labours such as surface finishes to concrete, but the method should prevent any major items being overlooked.

## The specification

If a specification is supplied, it may be read through cursorily first, not with the idea of mastering it in detail, but rather with a view to getting a general idea of its structure and content. A more detailed study may then be carried out of the sections relating to the element to be measured. For example, the excavation, concrete work, and brickwork may be read before beginning the taking-off for the substructure. It is useful when the taking-off is well advanced to go through the specification and run through in pencil all parts that have been dealt with but not paragraphs that will form preambles to the bill or that contain descriptions that are not repeated in the dimensions. If this is done, it is unlikely that anything specified will be missed. Under the Joint Contracts Tribunal (JCT) form of contract, the specification is not a contract document; however, should the bill and specification both form part of the contract, they will have to be integrated. In such cases, the wording of the bill descriptions can be reduced by reference to the specification, particularly in a coordinated document. When the specification is a contract document, the preambles and the preliminaries in the bill can be similarly reduced in length to avoid duplication.

More often than not, however, no formal specification is available at the outset. Brief specification notes may be supplied, and further notes must be made from verbal instructions given by the architect; these will form the nucleus of the specification, to be added to from time to time as queries are raised.

## Sequence of measurement

It is advisable when measuring to follow the same sequence in different parts of the work. For example, in collecting up the girth of the external walls of a building, it is a good idea to work clockwise, starting from, say, the top left-hand corner of the drawing. If this is done consistently, it will assist in reference later, when perhaps the length of a particular section of wall has to be extracted from a long collection. If a particular sequence of rooms has been adopted in measuring ceilings, the same sequence should be used for wall finish, skirtings, floors, etc. In this way, if all the finishes of a particular room are to be traced, it will be known in which part of each group the relative dimensions are to be found.

## Measurement of waste

It is normally stated as a general rule (though, like all rules, this has exceptions) that measurements of work are made to ascertain the net quantities as fixed or erected in the finished building. Wastage of material is allowed for by the contractor in the prices. This terminology should not be confused with the use of side cast or 'waste' calculations to establish the figures to be inserted into the dimension column.

## Overall measurements

It is usual in measuring to ignore in the first instance openings, recesses, and other features that can be dealt with by adjustment later. Brickwork, for instance, is measured as if there were no openings at all, and deductions are made when the windows, doors, or other openings are dealt with in the proper section. Plastering and similar finishes are measured in the same way. It simplifies the work to measure in this way. When, for instance, windows are being considered the sizes will be to hand, and openings will be measured to correspond. Moreover, it may happen that, say, internal plastering and windows are being dealt with by different takers-off, when it is obvious that the one who measures the windows is better able to make the deductions and adjustments. Most experienced takers-off decide that measuring overall with adjustment later is preferable to measuring net. This principle will also be found of value if, for example, a window should be forgotten, as the error would involve only the extra cost of the window over the wall and finishes, a much less serious matter than if nothing at all had been measured over the area of the window.

## Use of schedules

Early preparation of information schedules of such items as finishes, windows, and doors often exposes missing information and also provides a useful reference for the whole taking-off team. If like items are grouped together on the schedule, the taking-off process becomes more straight-forward and the need to continually refer to the specification during the taking-off is avoided.

## Use of scales

A warning should be given of the possibility of using the wrong scale in measuring from drawings. If possible, each side of the scale used should not have more than one variety of marking on each edge but this is not always practicable. The scale most easily available for general use is a standard metric scale having 1:5, 1:50, 1:10, 1:100, 1:20, 1:200, 1:250, and 1:2500 markings. Some surveyors prefer to have a separate scale for each variety, or one marked on one face only; but when working on two different drawings (e.g. 1:100 and 1:20) at the same time – as is quite common – it is very convenient to have both markings on the same scale. In any case, special care is necessary when measuring from different drawings to ensure that readings are taken from the correct markings of the scale. It may seem unnecessary to emphasise this but mistakes on this account are not unknown.

## Use of NRM2

The forms of contract agreed by the JCT provide that measurements shall be made in accordance with NRM2, and it is therefore of the utmost importance, where these forms of contract are used, that it should be followed. It should, nevertheless, be understood that, unless referred to in the contract, the Standard Method of Measurement (SMM) has no legal sanction and need not be adopted. The RICS describes NRM2 as a guidance note which means that it represents good practice found in industry. It may be deviated from but should any deviation result in a legal or disciplinary challenge, then justification for the deviation may be required.

## Decision on doubtful points

A thorough knowledge of NRM2 may still leave occasions when the taker-off must make decisions on the method of measurement or extent of descriptions for items not covered. When a rule of measurement is originated in such a way, it is often advisable to insert the method used as a preamble clause in the bill. When making such decisions, the taker-off should have one main consideration: what will best enable the estimator to understand (not merely guess) the work involved and enable it to be priced quickly and accurately?

## Descriptions

The requirements of NRM2 should be followed carefully when framing descriptions but it must be remembered that additional information should be given where necessary to convey the exact nature of the work to the estimator. Whilst it may be convenient to follow the order of the tabulated rules, this does not prevent the use of traditional prose in the framing of descriptions. Certain items such as waste of materials, square cutting, fitting or fixing materials or goods in position, plant, and other items are generally deemed to be included in descriptions. Such inclusions are listed at the start of each work section. Where NRM2 calls for a dimensioned description to be given, apart from the description of the item, all dimensions should be given to enable the shape of the item to be identified.

The taker-off must also be careful to see that the same wording is used when referring to the same item in different parts of the dimensions, as inconsistency in the descriptions may indicate that some distinction must be intended by the different phraseology. For instance, if the taker-off, describing plaster, writes '2 coat plaster on block walls', and then after several such items writes later '2 coat plaster on partitions', they may end up as different items in the bill, although this was not the intention of the taker-off. Therefore, when the same item appears in different places, it should be written in exactly the same form, or after the first time abbreviated with the letters 'a.b.' (for 'as before'; discussed further in the 'Abbreviations' section) to indicate that exactly the same meaning as before defined in NRM2. How PC and provisional sums are incorporated in a bill of quantities is dealt with in Chapter 20.

## Approximate quantities

Work to be carried out by the general contractor, which may be uncertain in extent, can also be provided for by means of approximate quantities (i.e. by measuring work in the normal way but keeping it separate in the bill and marking it 'approximate'). For instance, the foundations of a building, where the nature of the soil is uncertain, may be measured as shown on the drawings, and additional excavation, brickwork, and so on, measured separately and marked 'approximate' to cover any extra depth to which it may be necessary to take the foundations. It is thereby made clear that adjustment of these quantities on completion of the work is anticipated. Alternatively, if there is a considerable amount of such work, it may be contained in a bill of approximate quantities.

## Summary

1) Prepare a drawing register, and log any revisions received during bill preparation.
2) Check the plans, sections, and elevations for any clashes of information or missing details.
3) If measuring manually, put headings onto the dimension paper and number sheets. If using a computer system, structure the sections to be measured according to the structure of the final bill requirements.
4) Measure work from drawings in a set, logical pattern. One method frequently adopted is to start in the top left-hand corner of the drawing and work clockwise.
5) Write clearly and legibly, and space your work out.
6) Start by producing taking-off list for each section and building up a query (RFI) list as the take-off proceeds.
7) Measure items in the same sequence as your taking-off list, using figured dimensions in preference to scaling from the drawings.
8) Waste calculations should appear after descriptions.
9) Give locational notes against your dimensions so that work can be followed easily.
10) Use schedules where possible to save repetition.
11) Use 'to take notes' when you do not have sufficient information to measure an item but make sure to check these prior to final bill production.
12) When your measure is complete, go back over the drawings and line through all of the items measured. Then, repeat this process with your take-off list to ensure that you have not missed an item.
13) Check balances to ensure, for example, that all excavated material has been disposed of or used as filling where appropriate.
14) Stamp drawings as used for bill preparation if using actual drawings.

# 8

# Substructures

## Particulars of the site

Before beginning the measurement of substructures, the drawings must be examined to ascertain whether the existing ground levels are shown in sufficient detail for calculating average depths of excavation. If the levels are not shown or are insufficient, then it is necessary to take a grid of the levels over the site. Irrespective of taking levels, the surveyor should always visit the site to ascertain the nature and location of existing buildings, and details for preliminaries items and for the measurement of excavation work. Among items to be noted for the latter are vegetation to be cleared, the existence of topsoil or turf to be preserved, pavings or existing structures in the ground to be broken up and, if trial holes have been dug, the nature of the ground and the groundwater level. Where the proposed work consists mainly of alterations, an early visit to the site will be essential, and most of the taking-off may even have to be carried out there.

## Bulking

When measuring excavation, disposal, and filling, the dimensions are taken as in the ground, trenches being measured along their centre line and then multiplied by their width and depth as shown on the drawing. Soil increases in bulk when it is excavated but no account is taken for this in the bill of quantities, with the estimator having to make the due allowance during pricing.

## Removing topsoil

Where new buildings are to be erected on natural ground, it is necessary to measure a separate superficial item for the stripping of the vegetable soil or topsoil where it is to be preserved. This is measured over the area of the whole building including the projection of concrete foundations beyond external walls. A separate cubic item has to be taken for the disposal of the topsoil, giving the location. This would normally be on site for later reuse.

*Willis's Elements of Quantity Surveying,* Fourteenth Edition. Roy Hills and Sandra Lee.
© 2024 John Wiley & Sons Ltd. Published 2024 by John Wiley & Sons Ltd.

Any further excavation for trenches, basement, and so on, would then be measured from the underside of such topsoil excavation. If there are existing paths, paving, and the like over portions of the area to be stripped, an item must be taken for breaking out existing hard pavings, as a superficial item stating the thickness and the material, and measuring the removal as a separate item. Breaking out may be taken as extra over the excavation.

If the topsoil is to be replaced around the perimeter of the building, then this is measured in square metres stating the thickness.

Where the site is covered by existing buildings, the pulling down is measured as an item describing the size and construction of the building, no item for stripping topsoil being necessary. Demolition is usually taken to the existing ground level; breaking out below ground level is measured as a cubic item which may be taken as extra over the excavation. Work Section 3 contains detailed rules of measurement for demolition works.

Where the site is covered with good turf which is worth preserving, an item should be taken for lifting it (NRM2 5.5.1.1) and a separate item for relaying any turf to be reused (NRM2 37.4.1).

## Bulk excavation

Where a site is sloping, it is often more economical to set the ground-floor level so that one end of the site must be excavated into, that is, the underside of part of the hardcore bed will be below the level of the ground after the topsoil is stripped. Where this is the case, a cubic measurement is made of the excavating to reduce levels necessary from the underside of the topsoil excavation already measured to the underside of the hardcore bed. The depth for this item must be averaged, and it will generally be found that excavation is only necessary over part of the site, the level of the remainder being made up with hardcore or other filling. In Figure 10, it can be seen that the ground level must be reduced to a formation level of 44.00 (300 mm below the top of the floor slab). The contour of 44.00, known as the *cut and fill line*, is plotted on the plan as accurately as possible from the levels given; the area on the right-hand side of this contour is measured for excavation, and that on the left-hand side for filling.

Additional contours may have to be plotted to classify the excavation according to the maximum depths required by NRM2. It may, however, be more sensible to find an average depth for the whole excavation for measurement purposes and classify the depth in the description as the maximum on site. If this method is used, a statement should be made in the bill to this effect.

The bulk excavation to reduce levels must be measured before any foundation excavation, as it brings the surface to the 'reduced or formation level' from which the foundation excavation is measured, and, like the stripping of topsoil, it must be measured to the extreme projection of concrete foundations. A separate cubic item for disposal of excavated material must also be taken. It will often be necessary, where the floor level of the building is below the ground outside, to slope off the excavation away from the building, and possibly to have a space around the building excavated to below the floor level. In such circumstances, the additional excavation would be measured with the external works section.

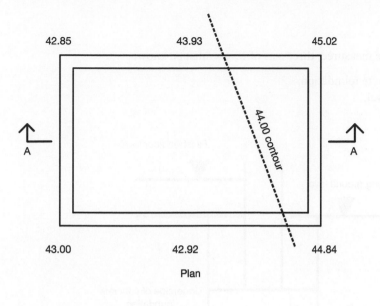

42.85          43.93          45.02

44.00 contour

A          A

43.00          42.92          44.84

Plan

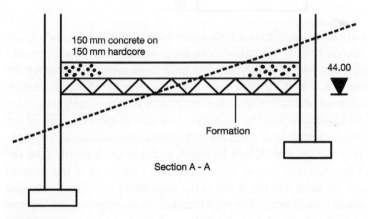

150 mm concrete on
150 mm hardcore

44.00

Formation

Section A - A

**Fig. 10**

## Excavation for paths

Stripping of topsoil and bulk excavation to reduce levels may also be required for paths, paved areas, and so on, around the outside of a building. External work of this nature is best measured all together after the building has been dealt with, as it usually forms a separate section in the bill. When paths about the building (thus overlapping the projection of foundations), the whole excavation necessary for the erection of the building should be measured with the building, the extra width necessary for the paths only being measured with the paths.

## Levels

Before foundations are measured, three sets of levels must be known:

1) Underside of concrete foundation.
2) Existing ground level.
3) Floor level.

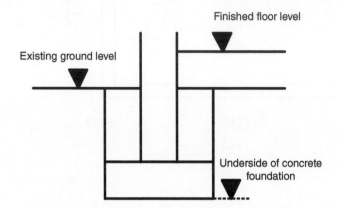

(1) and (2) are necessary to measure trench foundation excavation, and (1) and (3) are necessary to calculate correct heights of brickwork or other walling. The natural ground level, as has been pointed out, will probably vary and have to be averaged either for the whole building or for sections of it; if the floor level and the underside of the foundation are constant, the measurement of trench foundation excavation is fairly simple. However, both the underside of the foundation and the floor level may vary in different parts of the building, there being steps at each break in level, and sometimes the measurement of foundations thus becomes very complicated. It will be found useful to mark on the plan the existing ground levels at the corners of the building – if necessary, these can be interpolated from given levels – and in the same way the levels of the underside of the foundation can be marked on the foundation plan (if any). If stepped foundations are required, it will prove helpful if the foundation plan is hatched with distinctive colours to represent the varying depths of the underside of the foundations. If no foundation plan is supplied, the outlines of foundations can be superimposed on the plan of the lowest floor.

## Foundation excavation

In the simplest type of building, a calculation is made of the mean length of the foundation trench, as described in Chapter 6. Trench excavation is measured as a cubic item, and a separate cubic item of disposal of the excavated material must be taken. This mean length of the trench will also be the mean length of the concrete foundations. The width of trench will be the width of the concrete foundation as shown on the sections or foundation plan. The depth of trench will, if the underside of concrete foundation is at one level, be the

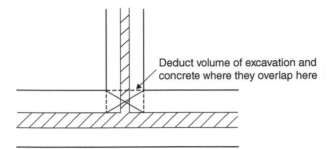

Deduct volume of excavation and concrete where they overlap here

**Fig. 11**

difference between that level and the average level of the ground, after making allowance for stripping of the topsoil or bulk excavation to reduce levels already measured. Where the underside of concrete is at different levels, theoretically the excavation for each section of foundation between steps should be measured separately, the lengths when measured being collected and checked with the total length ascertained. However, it may be found in practice that, the steps in the bottom of the trench are small and the ground falls in the same direction, so an average depth can be determined for larger sections of the building, if not for the whole. The depth of trench foundation excavation is given in 2 m stages. Alternatively, as mentioned in this chapter, in bulk excavation the maximum depth may be given and a statement made in the bill giving the method used.

A deduction should be made in the length of an internal wall foundation where it abuts against an external wall, by deducting from the length of the internal wall the projection of concrete foundation to the external wall at this point. A similar adjustment should be made where internal walls intersect. The necessity for this is best shown diagrammatically (Figure 11).

It will be seen that if the foundation for the internal wall were measured the same length as the wall (i.e. as dotted in Figure 11), the area marked with a cross would be measured twice for the excavation and concrete. Support to faces of excavations should also be considered at wall intersections to prevent overmeasure.

The amount involved being comparatively small, some surveyors may ignore the deduction but to do so when there are many intersections is to take an unnecessary and definite over measurement without any justifiable reason. Students often find it difficult to decide whether the maximum depths for trench excavation and the like should be calculated from the original ground level or from the level of the ground after topsoil has been stripped, where this is measured separately. It is usual to measure from the latter level as this is the commencing level of the actual excavation being measured. NRM2 requires that the commencing level of excavation must be stated where this is not the existing ground level, the depth classification of excavation being given from the commencing level.

## Earthwork support

Earthwork support is measurable under NRM2 as support to faces of excavation.

The wall between ground level and the d.p.c. is often constructed of facing bricks, which may be continued for one or two courses below ground level to allow for irregularities in the surface of the ground. The measurement of walling in substructure follows the rules for the measurement of general walling, covered in Chapter 9.

## Damp-proof courses

These are typically measured in metres if the width is less than or equal to 300 mm. This is taken as the width of the wall. If the d.p.c. is over 300 mm wide, then it is measured in square metres. Horizontal, raking, vertical, and stepped d.p.c.s are each kept separately. Pointing to the exposed edges is deemed to be included, but the thickness of the material, the number of layers, and the nature of the bedding material have to be stated. In the measurement, no allowance is made for laps.

When constructing basements, it is common to use asphalt damp proofing and tanking. These are measured using the area in contact with the base in square metres if the covering exceeds 500 mm. If less than or equal to 500 mm wide, then the units of measurement would be metres, The thickness, number of coats, and nature of the base and any surface treatment have to be included in the description together with the pitch. Internal angle fillets are measured separately in metres. Raking out joints of brickwork for a key is deemed to be included with the brickwork. No deduction is made for voids in asphalting less than or equal to $1\,m^2$.

Two examples are included here to demonstrate how the measurement rules are applied: for a simple wall foundation in Example 1 and for a more complex basement construction in Example 2.

## Approach taken to measurement in Example 1

Additional explanation is included here describing the approach taken to the measurement of the first example in this text. Understanding the process of construction for this simple building is essential. The ground is taken to be flat to simplify the example.

For the topsoil excavation the extreme length by breath is required, i.e. the building dimensions plus the spread each side. This is calculated in the waste or side-cast calculations and is shown by the red arrows on the following diagram.

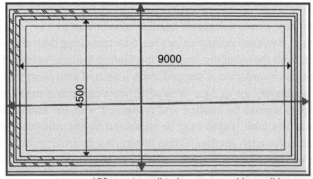

150 mm topsoil to be preserved in spoil heaps

**Plan**                                                                 **Scale 1:100**

The trench excavation and the subsequent disposal is measured next in cubic metres, along its centre line with the depth taken from the underside of the topsoil excavation down to the underside of the concrete foundation.

The diagram used here is taken from Example 1.

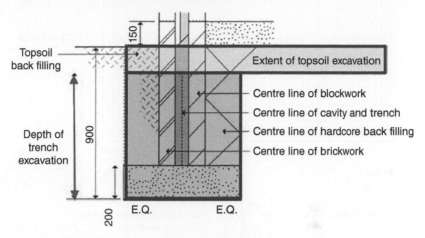

The concrete to be placed in the trench would only have the same centre line as the concrete if the cavity wall skins are the same width. The centre line of the concrete, multiplied by its width and depth will produce the volume in m³. Depending on the wall construction, this would then be measured for the height from the top of the concrete to d.p.c. level. By working through the example slowly the waste calculation should show how the dimensions have been derived.

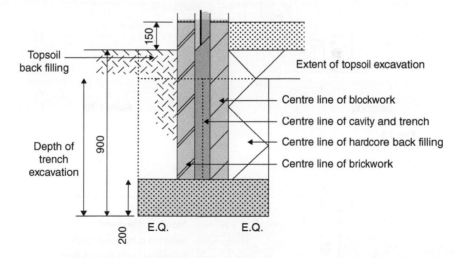

**Example 1   Wall Foundations**

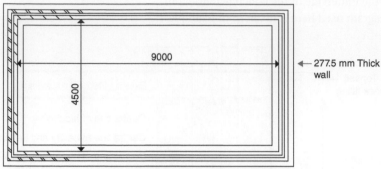

9000

4500

277.5 mm Thick wall

150 mm Topsoil to be preserved in spoil heaps

**PLAN**                                   **Scale 1:100**

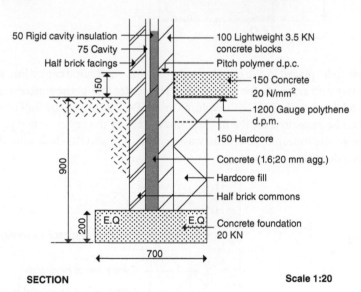

50 Rigid cavity insulation
75 Cavity
Half brick facings

150

900

200

E.Q.                    E.Q.

700

100 Lightweight 3.5 KN concrete blocks
Pitch polymer d.p.c.
150 Concrete 20 N/mm$^2$
1200 Gauge polythene d.p.m.
150 Hardcore
Concrete (1.6;20 mm agg.)
Hardcore fill
Half brick commons
Concrete foundation 20 KN

**SECTION**                                **Scale 1:20**

**Fig. 13**

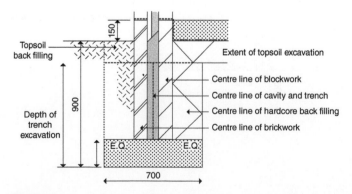

150

Topsoil back filling

900

Depth of trench excavation

E.Q.          E.Q.

700

Extent of topsoil excavation
Centre line of blockwork
Centre line of cavity and trench
Centre line of hardcore back filling
Centre line of brickwork

**Fig. 14**

Example 1
Wall foundations
Drawing: Figure 13 - no revision                    Teaching Notes

Taking off list
Topsoil excavation
Retaining topsoil
Foundation excavation
Disposal of subsoil
Disposal of groundwater
Earthwork support
Concrete foundation
Brick skin
Forming cavity
Block skin
Cavity fill
Backfill
Subsoil disposal adjustment
Damp-proof course
Hardcore bed
Damp-proof membrane horizontal
Damp-proof membrane skirting
Concrete slab

Request for Information (RFI) - notes        The RFI notes can be
Topsoil to temporary spoil heap?            later transferred to RFI
Original groundwater level?                 sheets and written up
DPM skirting detail?                        in full for distribution to
Blinding bed thickness?                     the relevant consultant.

| | Cavity wall thickness | |
|---|---|---|
| Half brick wall | 0.103 | Actual size 102.5 mm |
| Cavity | 0.075 | but rounds to 0.103 m. |
| Block skin | 0.100 | |
| | 0.278 | |

| | spread |
|---|---|
| trench width | 0.700 |
| Ddt cavity wall | 0.278 |
| 0.5 / | 0.422 |
| | 0.211 |

| | Overall length |
|---|---|
| | 9.000 |
| Add cavity wall b.s. | 0.556 |
| Add spread b.s | 0.422 |
| | 9.978 |

| | Overall width |
|---|---|
| | 4.500 |
| Add cavity wall b.s. | 0.556 |
| Add spread b.s. | 0.422 |
| | 5.478 |

(*Continued*)

*(Continued)*

| | | |
|---|---|---|
| 9.98<br>5.48 | 54.66 | Site preparation, remove topsoil average 150 mm deep<br>5.5.2.0.0 |
| 9.98<br>5.48<br>0.15 | 8.20 | Retaining excavated material on site, topsoil, to temporary spoil heaps<br>5.10.1.1.0 |

|  |  |
|---|---|
| Trench depth | |
|  | 0.900 |
| Ddt topsoil | 0.150 |
|  | 0.750 |

|  |  |
|---|---|
| Centre line of trench | |
|  | 9.000 |
|  | 4.500 |
| 2/ | 13.500 |
| Add corner adjustment | 27.000 |
| 4/ 2/ 0.5/ 0.278 | 1.112 |
|  | 28.112 |

| | | |
|---|---|---|
| 28.11<br>0.70<br>0.75 | 14.76 | Excavation, starting 150 mm below original ground level, foundation excavation, not exceeding 2 m deep<br>5.6.2.1.0<br>&<br>Disposal, excavated material off site<br>5.9.2.0.0 |
| Item | | Disposal, groundwater, depth below original ground level TBC.<br>5.9.1.1.0 |
| 25.31<br>0.75<br>30.91<br>0.90 | 18.98<br>27.82<br>46.80 | Support to faces of excavation, maximum depth 1 m, to trench excavation, distance between opposing faces less than 2 m<br>5.8.1.1.3 |

Earthwork support is measurable to all excavation faces over 250 mm deep.

|  |  |
|---|---|
| Centre lines | |
| Internal wall perimeter ab | 27.000 |
| Ddt corner adjustment | |
| 4/ 2/ 0.211 | 1.688 |
| Mean girth of internal support | 25.312 |
| Add corner adjustment | |
| 4/ 2/ 0.700 | 5.600 |
| Mean girth of external support | 30.912 |

| | | | |
|---|---|---|---|
| | | The following in-situ concrete works in substructures and not cast into formwork. | |
| 28.11 0.70 0.20 | 3.94 | Plain in-situ concrete, horizontal work, ≤ 300 mm thick, in structures, poured on or against earth or unblinded hardcore 11.2.1.2.1 | |
| 28.81 0.85 | 24.49 | Walls, overall thickness 0.103 mm, common brickwork, skins of holllow walls 14.1.1.1.0 | |

Height of brickwork
| | |
|---|---|
| Trench depth | 0.900 |
| Add to dpc | 0.150 |
| | 1.050 |
| Ddt conc fdn | 0.200 |
| | 0.850 |

Masonry centre lines
| | |
|---|---|
| Internal wall perimeter ab | 27.000 |
| Add corner adjustment 4/2/0.5/0.100 | 0.400 |
| CL internal block skin | 27.400 |
| Add ab | 0.400 |
| | 27.800 |
| Add corner adjustment 4/2/0.5/0.075 | 0.300 |
| CL of cavity | 28.100 |
| Add ab | 0.300 |
| | 28.400 |
| Add corner adjustment 4/2/0.5/0.103 | 0.412 |
| CL external brick skin | 28.812 |

| | | | |
|---|---|---|---|
| 28.81 0.23 | 6.48 | Ddt Walls, common brick abd & Walls, overall thickness 0.103 m, facing brick, skins of hollow walls 14.1.1.1.0 | Usually courses of facing brick are allowed for below the dpc i.e. 3 x 0.075 = 225 mm |

*(Continued)*

(*Continued*)

| | | | |
|---|---|---|---|
| 27.40<br>0.85 | 23.29 | Walls, overall thickness 0.100 m, blockwork, skins of hollow walls<br><br>14.1.2.1.0 | |
| 28.10<br>0.85 | 23.89 | Forming cavity, width 0.075 m, using wall ties<br><br>14.14.1.1.0 | Specification will also contain details of wall ties. |
| 28.10<br>0.08<br>0.70 | 1.48 | Plain in-situ concrete, vertical work, > 300 mm thick, in structures<br><br>11.5.2.1.0 | Weak concrete cavity fill |

Height | 
--- | ---
 | 0.900
Ddt conc fdn | 0.200
 | 0.700

| | | | |
|---|---|---|---|
| 27.40<br>28.81 | 27.40<br>28.81<br>56.21 | Damp proof courses ≤ 300 mm wide, single layer, horizontal<br><br>14.16.2.3.0 | |
| 30.07<br>0.21<br>0.55 | 3.49 | Filling obtained from excavated material, final thickness of filling exceeding 500 mm deep, average 150 mm layers<br><br>5.11.2.2.0 | |

Backfill depth

| | |
|---|---|
| depth of backfill | 0.750 |
| Ddt conc fdn | 0.200 |
| | 0.550 |

Backfill CL

| | |
|---|---|
| CL external brick skin ab | 28.812 |
| Add corner adjustment | |
| 4/2/0.5/0.103 | 0.412 |
| External perimeter of brick skin | 29.224 |
| Add corner adjustment | |
| 4/2/0.5/0.211 | 0.844 |
| CL of backfill | 30.068 |

&

Ddt
Disposal offsite abd

| | | | |
|---|---|---|---|
| 30.07<br>0.21 | 6.34 | Filling obtained from excavated material, final thickness of filling, not exceeding 500 mm deep, finished thickness 150 mm<br><br>5.11.1.2.0 | topsoil backfill |

| | | | |
|---|---|---|---|
| 26.16<br>0.21<br>0.55 | 3.04 | Imported filling, beds and voids exceeding 500 mm, level, average depth of layers 150 mm<br><br>5.12.3.1.2 | |

<u>Hardcore fill depth</u>

<u>Ddt</u> conc fdn    0.900<br>0.200<br>0.700

<u>Ddt</u> hardcore slab    0.150<br>0.550

<u>Centre lines</u>

Internal wall perimeter ab    27.000<br>
<u>Ddt</u> corner adjustment<br>
4/2/0.5/   0.211    0.844<br>
26.156

| | | | |
|---|---|---|---|
| 9.00<br>4.50<br>0.15 | 6.08 | Imported filling, beds and voids over 50 mm thick but not exceeding 500 mm thick, finished thickness 150 mm, level, average depth of layers 150 mm<br><br>5.12.2.1.2 | |
| 9.00<br>4.50 | 40.50 | Damp-proof membrane, over 500 mm wide, 1200 gauge, horizontal<br><br>5.16.2.1.0 | |
| 2/ 9.00<br>2/ 4.50 | 18.00<br>9.00<br>27.00 | Damp-proof membrane, not exceeding 500 mm wide, 1200 gauge, vertical<br><br>5.16.1.3.0 | |
| 9.00<br>4.50<br>0.15 | 6.08 | Plain in-situ concrete, horizontal work, less than or equal to 300 mm thick, in structures<br><br>11.2.1.2.0 | |

*(Continued)*

| | | | |
|---|---|---|---|
| 4.90<br>4.90 | 24.03 | Asphalt tanking coverings, ><br>500 mm wide, horizontal, 30 mm<br>thick in 3 coats on concrete base<br>19.1.1.0.1 | |

<u>Overall dimensions</u>

4.000

1.5 bk wl b.s.

2/0.328  0.656

<u>    4.656</u>

asphalt 2 ct b.s.

2/0.020  0.040

<u>    4.696</u>

0.5 bk wl b.s.

2/0.103  0.206

<u>    4.902</u>

| | | | |
|---|---|---|---|
| 4.66<br>4.66<br>0.15 | 3.25 | Reinforced in situ concrete,<br>horizontal work, ≤ 300 mm thick,<br>in structures<br>11.2.1.2.0 | |

<u>Overall dimensions</u>

4.000

Add 1.5 brick wall b.s.

2/0.328  0.656

<u>    4.656</u>

| | | | |
|---|---|---|---|
| 4/  4.66 | 18.62 | Plain formwork, edges of<br>horizontal work, ≤ 500 mm high,<br>width 150 mm<br>11.15.1.0.0 | |

| | | | |
|---|---|---|---|
| 4.56<br>4.56 | 20.76 | Reinforcement, mesh, fabric<br>reference A142<br>11.37.1.0.0 | |

<u>Overall dimensions</u>

ab  4.656

<u>Ddt</u> concrete cover b.s.

2/0.050  0.100

<u>    4.556</u>

| | | | |
|---|---|---|---|
| 17.31<br>2.38 | 41.20 | Walls, overall thickness 328 mm, common brickwork<br>14.1.1.0.0<br><div align="right">height<br>2.400<br>Ddt asphalt 0.020<br>2.380</div> | |
| 18.62<br>2.40 | 44.70 | Asphalt tanking coverings > 500 mm wide, vertical, 20 mm thick in 2 coats on brickwork base<br>19.1.3.0.1 | Asphalt measured on area in contact with base so no centre line necessary. |
| 18.62 | 18.62 | Asphalt tanking coverings ≤ 500 mm wide, vertical, 20 mm thick in 2 coats on concrete base<br>19.1.3.0.1 | |
| 18.78 | 18.78 | Internal angle fillet, 30 × 30 mm, on asphalt base<br>19.11.1.0.1 | |
| 17.31 | 17.31 | Asphalt tanking coverings ≤ 500 mm wide, horizontal, 20 mm thick in 2 coats on brickwork base<br>19.2.1.0.2 | |
| 19.20<br>2.55 | 48.95 | Walls, overall thickness 103 mm, common brickwork, built against other work<br>14.1.1.4.0<br><div align="right">Height<br>2.400<br>Add conc slab 0.150<br>2.550</div> | |

(*Continued*)

# 9

# Walls

## Measurement of brickwork

Brickwork generally is measured in square metres, taking the mean girth of the wall (centre line) by the height. The heights of walls often vary within the same building, but it is usually possible to measure to some general level, such as the main eaves, and then to add for gables and so on, and deduct for lower eaves and the like. No deductions are made for voids or built-in items within the area of brickwork with a cross-sectional area less than $0.50\,m^2$.

In the case of internal walls and partitions, it is necessary to ascertain whether or not these go through floors and to measure the heights accordingly. If the thickness of a wall reduces, say, at floor level or at a parapet, then it should be remembered that the mean girth of the wall changes.

All openings in walls are usually ignored when measuring at this stage, the deduction for these openings being made when the doors or windows are measured, as described in Chapter 13.

Refer to the approach taken to the measurement of walls in Example 3.

## Subdivision

The measurement of walls can be conveniently divided into the following subdivisions:

- External walls
- Internal walls and partitions
- Projections of piers and chimney breasts
- Flues and chimney stacks.

In each of the brickwork classifications, brickwork is described as skins of hollow walls, battered, tapered one side or both sides, built against other work or used as formwork. Isolated piers are defined as such when the length of the pier is less than four times its thickness, except where caused by openings.

*Willis's Elements of Quantity Surveying,* Fourteenth Edition. Roy Hills and Sandra Lee.
© 2024 John Wiley & Sons Ltd. Published 2024 by John Wiley & Sons Ltd.

## Measurement of projections

Projections are defined as attached piers (if their length on plan is less than four times their thickness) plus plinths, oversailing courses, and other similar items. Projections are measured linear, and the width and depth of the projection are given in the description. Horizontal, raking, and vertical projections are kept separately. Example 4 at the end of this chapter gives sample measures for these.

## Descriptions

The information that should be given, in addition to the matters discussed in this chapter, includes the kind, quality, and size of the bricks; the bond, composition, and mix of mortar; and the type of pointing. These items, particularly if common to all the work, can be included in the bill as preambles or headings to reduce the length of descriptions. It should be remembered that, as mentioned here, mortar mixes may vary and, apart from substructure work, brickwork above eaves level in parapets and chimney stacks may well be specified to be in cement mortar.

## Brickwork thickness

As the thickness of the brickwork has to be given in the description, it is usually more convenient to give this in relation to the number of bricks.

The generally accepted average size of a brick is $215 \times 102.5 \times 65$ mm, which with a 10 mm mortar joint becomes $225 \times 112.5 \times 75$ mm. This gives the following wall thicknesses:

| | |
|---|---|
| • Half brick | 102.5 mm |
| • One brick | 215 mm |
| • One and a half brick | 327.5 mm |
| • Two bricks | 440 mm. |

## Common and facing brickwork

When describing brickwork, it is necessary to describe the finish. Common brickwork is walling in ordinary or stock bricks without any special finish and usually hidden from view. If the brickwork is exposed to view and finished with a neat or fair face, then the finish should be described as facing, either one side or both sides. Walls one brick thick and over may have facing bricks on one side and common bricks on the other, this being a more economical way of constructing the wall if only one face shows.

## Cutting, grooves, and the like

All rough cutting on common brickwork and fair cutting on facework is deemed to be included. Rough and fair grooves, throats, mortises, chases, rebates, holes, stops, and mitres are all also deemed to be included.

## Returns and reveals

The labour in forming returns at reveals at the ends of walls is deemed included, and therefore nothing has to be measured for these.

## Hollow (cavity) walls

Each skin of a hollow wall is measured along its centre line in square metres and described as skins of hollow walls. Forming the cavity is measured in square metres, stating the width of the cavity.

The type and spacing of the wall ties have to be given in the description of forming the cavity. If rigid sheet insulation is required in the cavity, this is also included in the description, stating the type, thickness, and fixing method. Foam, fibre, or bead cavity filling is measured separately as a superficial item, stating the width of the cavity, the type and quality of the material, and the method of application. Filling cavities with concrete below ground level is measured as a cubic item in Work Section 11 and described as vertical work, stating the thickness in the description as less than or equal to 300 mm or as exceeding 300 mm. Closing cavities is described as extra over for wall perimeters and measured in linear metres, stating the width of the cavity and the method of closing. Cavity closers may be enumerated as proprietary and spot items.

## Bands

These are items such as brick on edge or end bands, basket pattern bands, moulded, or splayed plinth cappings, moulded string courses, moulded cornices, and the like. Horizontal, raking, vertical, and curved bands are measured as separate linear items, stating the width and classified as follows:

- Flush
- Sunk (depth of set back stated)
- Projecting (depth of set forward stated).

If the bands are constructed entirely of stretchers or entirely of headers, this has to be stated, and if curved, the mean radius given. Ends and angles on the bands are deemed included.

### Example 3  Walls and Partitions (Above damp-proof course)

### Approach to measurement

The walls are measured as though they formed a square box and an adjustment for the splayed walls then follows. The internal girth is found and then using the four times twice times the distance moved principle, the centre line of each skin of the hollow wall are established. The area requires the centre line to be multiplied by the height from the damp-proof course to the top of the wall (Figure 17).

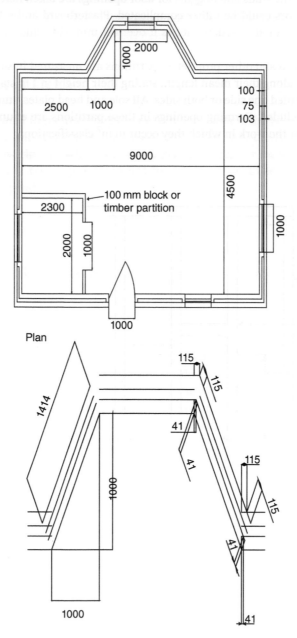

Plan

Plan of bay window

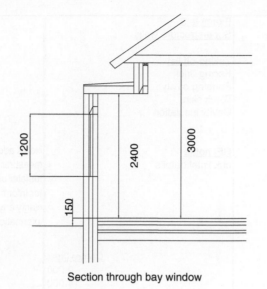

1200
2400
3000
150

Section through bay window

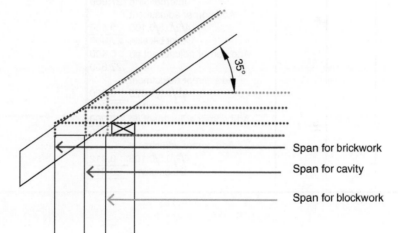

35°

Span for brickwork
Span for cavity
Span for blockwork

**Fig. 17**

Example 3
Superstructure walls

Taking-off list
Facing brick
Forming cavity
Block skin
Cavity insulation


RFI notes
chk lintel details

|  | Notes added as measurement progresses for later use in Request for Information sheets to clarify any missing design information. |

Centre lines
9.000
4.500
2 / 13.500
internal girth 27.000
Add corner adjustment
4/2/0.5/0.100   0.400
CL int blk skin 27.400
Add corner adjustment ab   0.400
27.800
Add corner adjustment
4/2/0.5/0.075   0.300
CL cavity = 28.100
Add corner adjustment ab   0.300
28.400
Add corner adjustment
4/2/0.5/0.103   0.412
CL bk skin = 28.812

| | | | ALTERNATIVE CENTRE LINE CALCULATION | |
|---|---|---|---|---|
| | | | internal girth ab   27.000<br>Add corner adjustment<br>4/2/0.5/0.100   0.400<br>CL int blk skin =  27.400 | |
| | | | internal girth ab   27.000<br>Add corner adjustment<br>4/2/0.138   1.100<br>CL cavity =  28.100 | block + half the cavity |
| | | | internal girth ab   27.000<br>Add corner adjustment<br>4/2/0.138   1.812<br>CL bk skin =  28.812 | block + cavity + half brick |
| 27.40<br>2.95 | | 80.83 | Walls, overall thickness 100 mm,<br>blockwork, skins of hollow walls<br>14.1.2.1.0      <u>Height</u><br>                 3.000<br>   <u>ddt</u> wallplate   <u>0.050</u><br>                 2.950 | |
| 28.10<br>3.00 | | 84.30 | Forming cavity, width 75 mm<br>including 50 mm thick insulation<br>14.14.1.0.0 | Note: forming cavity and insulation may be measured together as a composite item. |
| 28.81<br>3.00 | | 86.44 | Walls, overall thickness 103 mm,<br>facing brickwork, skins of hollow<br>walls<br>14.1.1.1.0 | |

*(Continued)*

(*Continued*)

| | | | |
|---|---|---|---|
| | | Adjustment for bay window. | |
| 4.00 2.40 | 9.60 | <u>ddt</u> blk wl sohw ab | |
| | | Bay <u>length on plan</u> | |
| | | & 1.000 | |
| | | 2.000 | |
| | | <u>ddt</u> 1.000 | |
| | | forming cavity ab 4.000 | |
| | | & | |
| | | <u>ddt</u> bk wl sohw ab | |
| | | blk wl sohw ab | |
| 4.83 2.40 | 11.59 | <u>Actual bay length</u> Slope length b.s. 2.828 | Using Pythagoras, the |
| | | Add bay width 2.000 | slope spreadsheet |
| | | 4.828 | formula is :- |
| | | & | =SQRT((1.00^2)+(1.00^2)) |
| | | forming cavity ab | |
| | | & | |
| | | bk wl sohw ab | |

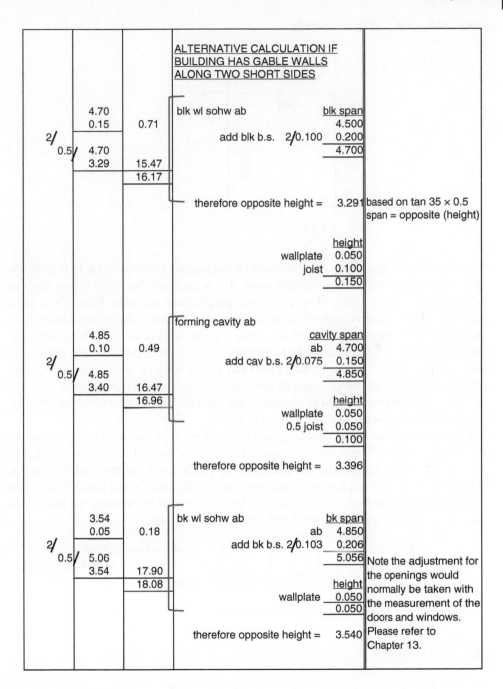

ALTERNATIVE CALCULATION IF
BUILDING HAS GABLE WALLS
ALONG TWO SHORT SIDES

|   |   |   |   |
|---|---|---|---|
| | 4.70 | | blk wl sohw ab                    blk span |
| | 0.15 | 0.71 |                                              4.500 |
| 2/ | | |                 add blk b.s.  2/0.100   0.200 |
| 0.5/ | 4.70 | |                                              4.700 |
| | 3.29 | 15.47 | |
| | | 16.17 | |

therefore opposite height =   3.291   based on tan 35 × 0.5
                                                        span = opposite (height)

                                                               height
                                              wallplate   0.050
                                                    joist   0.100
                                                               0.150

forming cavity ab

|   |   |   |   |
|---|---|---|---|
| | 4.85 | | forming cavity ab                cavity span |
| | 0.10 | 0.49 |                                        ab   4.700 |
| 2/ | | |                 add cav b.s. 2/0.075   0.150 |
| 0.5/ | 4.85 | |                                              4.850 |
| | 3.40 | 16.47 | |
| | | 16.96 | |

                                                               height
                                              wallplate   0.050
                                              0.5 joist   0.050
                                                               0.100

therefore opposite height =   3.396

|   |   |   |   |
|---|---|---|---|
| | 3.54 | | bk wl sohw ab                       bk span |
| | 0.05 | 0.18 |                                        ab   4.850 |
| 2/ | | |                 add bk b.s. 2/0.103   0.206 |
| 0.5/ | 5.06 | |                                              5.056 |
| | 3.54 | 17.90 | |
| | | 18.08 | |

Note the adjustment for
the openings would
normally be taken with
                                                               height     the measurement of the
                                              wallplate   0.050   doors and windows.
                                                               0.050     Please refer to
                                                                             Chapter 13.

therefore opposite height =   3.540

**Alternative stud partition example**

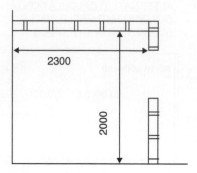

**Fig. 18**

The following example shows the dimensions if the internal partitions were built in block-work and then an alternative measure given if it is assumed that 100×50 mm studs at 400 mm centres will be used. Head plate and sole plate are also to be 100×50 mm sawn softwood. There would be studs fixed vertically at the junction with the inner block skin. Additional studs are then positioned at corners and openings to form the main structure. These might be a larger size than the basic studs or may even be double studs. Infill studs are then spaced equally between wall plates and corner studs or those at openings, at no more than 400 mm centres (Figure 18). Horizontal noggins are then required to brace the wall; in a 3 m high wall, it would be normal to have one line of horizontal members although this would depend on the individual drawing details.

Adjustments for the door opening should be made after the members have been meas-ured. To decide how many studs are required, the length of the wall is taken and divided by the spacings. The result must be rounded up to the nearest whole number and then, as this is the number of spacings, it is necessary to add one to allow for the end stud. On plan, this would be depicted as below. The same principle is also used to calculate the number of joists required in a floor or the number of roof trusses required.

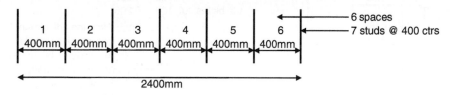

| | | | |
|---|---|---|---|
| | 4.40 | | Internal wall |
| | 3.00 | 13.20 | Walls, overall thickness 100 mm, blockwork 14.1.2.0.0 |

CL
2.300
2.000
4.300

add corner adjustment
1/2/0.5/0.100  0.100
4.400

ALTERNATIVE MEASUREMENT
FOR TIMBER STUD PARTITION

studs

Primary or structural timbers, nominal size 50×100 mm, partition and wall members, fixing centres 400 mm

| | | | |
|---|---|---|---|
| 7/ | 2.90 | 20.30 | 16.1.1.7.4 |
| 6/ | 2.90 | 17.40 | |
| | 2.90 | 2.90 | (extra stud in corner |
| ddt | 2.15 | −2.15 | (stud at dr opg |
| | | 38.45 | |

wl lgths
2.300
add for wall  0.100
2.400
spacing  0.400
6
add to covert from spaces  1
7

2.000
spacing  0.400
5
add to covert from spaces  1
6

height
3.000
ddt headplate and soleplate
2/0.05  0.100
2.900

door opg ht
2.100
plus head  0.050
2.150

(*Continued*)

(*Continued*)

| | | | | | |
|---|---|---|---|---|---|
| | | | Diagram C | | |

| | | | | |
|---|---|---|---|---|
| 5.00<br>3.00 | 15.00 | | Walls, overall thickness 215 mm, brickwork, one side common bricks and one side facing brick, facing one side<br>14.1.1.0.0 | Assumed one side facing brick and one side common bricks. Wall length 5 m by 3 m high. |
| 3.00 | 3.00 | | Attached projection, facing brick, 327.5 × 102.5 mm, vertical<br>14.5.1.1.0 | Measured as attached projection as length on plan ≤ 4 times the thickness, i.e. 327.5 ≤ 410. |

Diagram D

| | | | | |
|---|---|---|---|---|
| 5.00<br>3.00 | 15.00 | | Walls, overall thickness 215 mm, brickwork, facing bricks, facing both sides<br>14.1.1.0.0 | Assumed built entirely of facings and wall length 5 m by 3 m high. |
| 3.00 | 3.00 | | Attached projections, facing brick, 327.5 x 102.5 mm, vertical<br>14.5.1.1.0 | Measured as attached projection as length on plan ≤ 4 times the thickness, i.e. 327.5 ≤ 410. |

| | | | | |
|---|---|---|---|---|
| 3.00 | 3.00 | **Diagram E**<br><br>327.5 ↕ 102.5<br><br>Isolated pier, 327.5 × 102.5 mm, common brick, vertical<br>14.4.1.1.0 | Measured as isolated pier as length on plan ≤ 4 times the thickness, i.e. 327.5 ≤ 410. | |
| 0.66<br>3.00 | | **Diagram F**<br><br>655 ↕ 102.5<br><br>Wall, overall thickness 102.5 mm, brickwork, common bricks, facing both sides<br>14.1.1.0.0 | Measured as wall because length on plan > 4 times the thickness, i.e. 656 > 410. | |

# 10

# Floors

## Timber sizes

Timber for use in construction is sawn into standard sectional dimensions, and the size thus created is known as the *nominal* or *basic* size. When the timber is processed, planed, or wrot, it is reduced in size, usually by about 2 mm on each face, and the resultant size is known as the *finished* size. The original size of processed timber is sometimes described as *ex* or *out of*, thus 'ex 100 × 25 mm' would be '96 × 21 mm finished'.

Regularising is a machine process by which structural timber is sawn to a uniform size (e.g. joists sawn to an even depth). For regularising, 3 mm should be allowed off the size of timber of up to 150, and 5 mm above this size.

NRM2 states that sizes of timber are deemed to be nominal unless otherwise described. Confusion may be avoided if the sizes given in the bill are as they are shown on the drawings, provided that these are consistent. Structural timber when not exposed is usually left with a sawn finish, and the sizes given on the drawing are nominal. When architects are designing joinery, which may have to be to precise dimensions, they often prefer to give finished sizes. Care must be taken to ascertain which sizes are given on the drawing and to make it clear in the bill if finished sizes are being used.

The amount of timber removed in creating a wrot face on timber is known as the *planing margin*; this should be given as a preamble in the bill. Reference may be made to the appropriate standard for various planing margins according to the sizes and the end use of the timber.

The specification for tongued and grooved floor boarding can cause confusion, as the finished width on face and the finished thickness may be quoted; for example, 90 × 21 finished boarding would be 100 × 25 nominal (sawn) size to allow for the tongue and planing margin.

Further consideration should be given to the sizes of timber readily available. For example, the width of a door lining to a 75 mm block partition with 13 mm plaster on both faces would require a finished lining width of 101 mm. The nearest nominal size of timber available is 115, or 125 mm for some types of timber. This information should be clarified by the designer.

*Willis's Elements of Quantity Surveying*, Fourteenth Edition. Roy Hills and Sandra Lee.
© 2024 John Wiley & Sons Ltd. Published 2024 by John Wiley & Sons Ltd.

Timber exceeding 5.1 m in length is usually more expensive; in the examples, an allowance in the length of timbers has been made for jointing where appropriate. NRM2 requires structural timbers over 6 m long in one continuous length to be described as such (16.1.1.1.9).

## Subdivision

The measurement of floors is conveniently divided into two main subdivisions of finishes (Work Section 28) and construction (Work Section 16). These subdivisions may be taken storey by storey, or the finishes may be taken first for the whole building, followed by the construction. Whether or not the finishes are taken before the construction is really a matter of personal choice, although measurement, and therefore knowledge, of the finishes may assist in deciding the form of construction to be used. For example, a timber upper floor may have a general finish of boarding, but a small tiled area may require a different form of construction which in any case is a matter for the designer to decide.

Frequently, floor finishes are measured with the internal finishes section, because measurements used for ceilings may apply to floors. This is, however, a matter for agreement between the surveyors measuring the two sections, although it is customary to measure the finish to timber floors with the construction.

## Timber floor construction

The measurement of timber floors may be divided as follows:

- Floor joists/or beams
- Hangers
- Strutting
- Ties to walls
- Insulation.

Timber-suspended ground floors are not usually cost effective but may be used to avoid excessive fill below solid floors or on sloping ground. The under-floor space should be ventilated; air bricks and sleeper walls built honeycombed with a damp-proof course if required, and they are measured with either floors or substructure but billed with the latter.

Timber wall plates are measured linear, and the length measured should not allow for any laps or joints required. Timber joists are measured in the same way but are described as floor joists.

To ascertain the number of joists in a room, take the length of the room at right angles to the span of the joists and subtract, say, a 25 mm clearance plus half the thickness of the joist at each end. This will give the distance from the centre lines of the first and last joists, which is divided by the spacing of the joists. The result, which should be rounded to the next whole number, gives the number of spaces between the joists. To this number, one must be added to give the number of joists rather than the spaces. Two points should be borne in mind before making the calculation. First, the room size used for the calculation should be that of the room below the floor as the joists will be supported by the walls of that lower room. Secondly, if any of the intermediate joists are to be in a fixed position, such as a trimmer around an opening, two separate dimensions should be taken for the division on either side of the fixed joists, thus avoiding any increase in the maximum spacing. Additional joists may be required to support upper-floor partitions, and care must be taken to include for these. Figure 20 shows how to calculate the number of joists required for a timber ground floor.

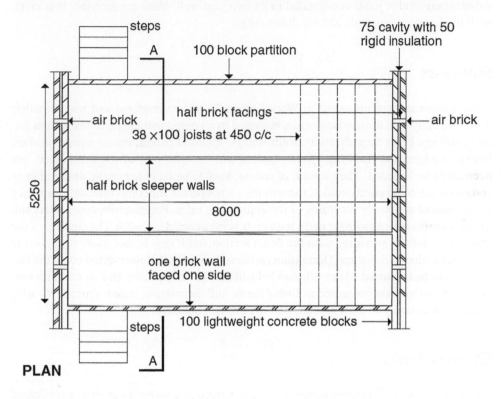

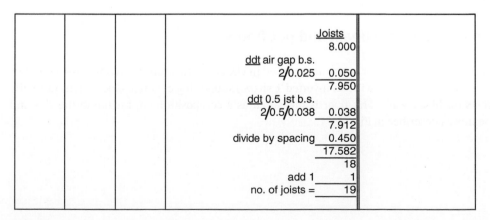

**Fig. 20**

The trimming of timber joists for staircases, ducts, hearths, access panels, and so on is usually taken with the floor construction. Joists used in trimming should be increased in thickness by 25 mm. Displaced joists and floor coverings have to be deducted.

Actions to prevent joists twisting, such as herringbone or solid strutting, which may not be shown on the drawings, should be taken. This is measured linear over the joists, and there should be one row, say, every 2.4 m. Building regulations may require ties to be provided where timber joists run parallel to an envelope wall; these are probably best taken with the floor construction. Refer to Example 5.

## Staircases

Timber staircases are measured in Work Section 25 and enumerated and may be either described fully with dimensions or accompanied by a component detail. Items such as linings, nosings, cover moulds, trims, soffit linings, spandrel panels, ironmongery, finishes, fixings, wedges, and the like, where supplied with or with part of the component, are deemed to be included. They would, of course, have to be included in the description or shown on the component detail. If these items are not part of the composite item, then they are measured separately according to the appropriate rules. For example, cover fillets and nosings are measured as linear items with ends being deemed included. The creation of the stairwell is usually measured with the floors section, but it may be necessary to adjust the internal finishes at this stage. Decoration on the staircase itself, if not carried out at the factory, has to be measured. Handrails and balustrades that are isolated and do not form part of a staircase unit are measured as linear items, and their size is stated. Ramps, wreaths, bends, and ornamental ends on handrails are deemed included.

## Concrete floors

The measurement of concrete upper floors is described in Chapter 14 as part of reinforced concrete structures.

## Precast concrete beam and pot floors

A common ground floor construction used in UK domestic properties comprises a series of precast concrete ribs (with an inverted T shape) supporting concrete blocks. This normally rests on block walls. The floor is measured as a composite item, including the ribs and beams as described in Figure 21.

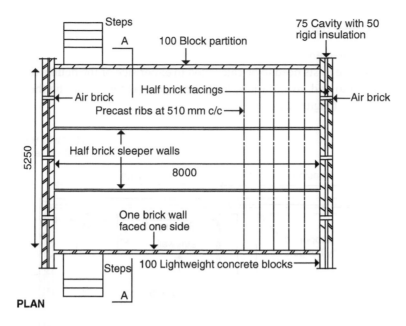

**PLAN**

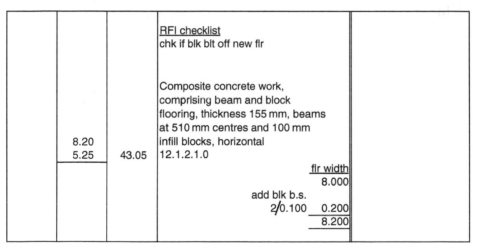

| | | | | | |
|---|---|---|---|---|---|
| | | | RFI checklist<br>chk if blk blt off new flr<br><br><br>Composite concrete work,<br>comprising beam and block<br>flooring, thickness 155 mm, beams<br>at 510 mm centres and 100 mm<br>infill blocks, horizontal<br>12.1.2.1.0 | | |
| 8.20<br>5.25 | | 43.05 | | | |

flr width 8.000
add blk b.s. 2/0.100 0.200
8.200

**Fig. 21**

being measured. The span inside the walls being 3000 mm, half the span for this purpose will be $1500 + 215 + 230 = 1945$ mm. The angle of pitch being 30°, the length of the slope will be:

$$1945 \div \text{cosine} 30$$

$$= 1945 \div 0.866$$

$$= 2246 \text{ mm}$$

and this can be checked by scaling or by interrogating the CAD drawing if available.

It should be noted that this calculation gives the length along the top of the rafter to the centre line of the ridge. This may have to be adjusted for the length of the covering, which may project into a gutter.

## Hips and valleys

The length of a hip or valley in a pitched roof must be calculated from a triangle, there usually being no true section through it from which it can be scaled. When the pitch of the hipped end is the same as that of the main roof, then the length of the hip may be found from half the span and the length of the roof slope using Pythagoras' theorem.

For example, in Figures 26 and 27:

$$\text{BE and BD} = \frac{1}{2}\text{span}$$

$$\text{AD and AE} = \text{length of slope}$$

$$\therefore \text{AB} = \sqrt{\text{AD}^2 + \text{BD}^2}$$

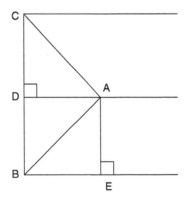

**Fig. 26**

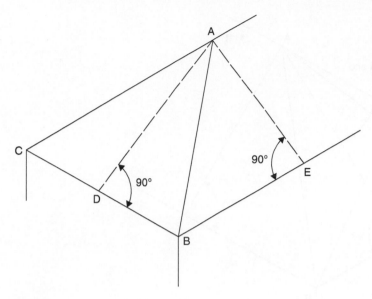

**Fig. 27**

## Broken-up roofs

It should be noted that when a roof is broken up by hips and valleys, so long as the pitch is constant the area will always be the overall length multiplied by twice the slope. The area of tiling on a roof hipped at both ends is therefore measured in the same way as if it were gabled, the only difference being that if it were gabled, the dimension of the length would probably be smaller, the projection of verges being less than that of eaves.

For example, in Figure 28:

Area of triangle ABD = area of triangle ABF

$$BE = BD = AF = \frac{1}{2} \text{ span}$$

AB is common

$$A\hat{D}B = 90° \text{ and } A\hat{F}B = 90°$$

With two sides and one angle equal, the triangles ABF and ABD are equal in area.

Just as the length of slope is the length on plan divided by the cosine of the angle of pitch, so the area of a roof of constant pitch is the area on plan divided by the cosine of the angle of pitch; this formula can be a useful check on the measurements when worked out.

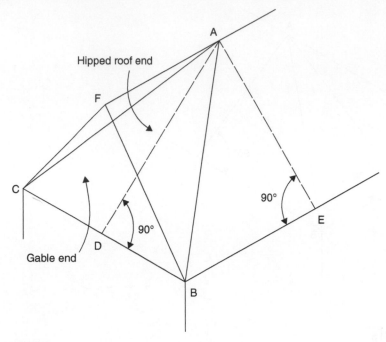

**Fig. 28**

## Trussed rafters

Trussed rafters are enumerated and described, including a manufacturer's reference if available. If not available, a dimensioned diagram is also a clear way to describe a truss. Bracing and binders to the construction are taken as linear items as note exceeding 600 mm wide. When calculating the number of trussed rafters, consideration should be given to the position of water storage tanks and the like, as their support may require additional trusses, platforms, and bracing. This will probably involve discussion with the surveyor measuring the plumbing installation; agreement should be reached on where additional items will be measured and by whom.

Trusses need longitudinal and diagonal bracing to ensure their stability and galvanised steel straps are used to tie the trusses to end gable walls.

## Tile or slate roof coverings

The first measurement in this type of roof should be the area of coverings; the description will include battens and underlay. If the area of one slope is entered in the dimensions, care must be taken to multiply this by two to allow for both slopes. Following the coverings, one should proceed around the edges of the roof, measuring abutments, eaves, verges, ridges, hips, and valleys as boundary work. The adjustments for chimneys and dormers and any appropriate metal gutters, soakers, and flashings could conclude the coverings measurement or be taken as a separate section at the end after the construction.

Unless otherwise specified, customary girths for leadwork to pitched roofs are as follows:

| | |
|---|---|
| • Cover flashings | 150 mm |
| • Stepped flashings (over soakers) | 200 mm |
| • Ditto (without soakers) | 300–350 mm |
| • Aprons | 300 mm |
| • Soakers 200 mm wide | length = gauge + lap + 25 mm<br>(number = length of slope ÷ gauge) |

## Eaves and verge finish

Verge, eaves, fascia, and barge boards are measured linear (in Work Section 16) if not exceeding 600 mm wide, giving their thickness in the description. Painting the boards should be taken at the same time in square metres (see Work Section 29), unless the member is isolated and ≤ 300 mm wide, when it is measured linear.

## Rainwater installation

Rainwater gutters are measured linear over all fittings, these being enumerated and described gutter ancillaries. Rainwater downpipes are measured in a similar manner. The description of gutters and rainwater pipes should include their size and method of fixing to the background. Rainwater outlets are enumerated and gratings may be included in the description.

A careful check should be made to ensure that all roof slopes are drained, that there are sufficient rainwater pipes, and that the drainage plan shows provision for each pipe. If the sizes of gutters and pipes are not shown, then these have to be calculated by the designer, but it should be remembered that a variety of sizes, although theoretically correct, does not always lead to an economic solution.

## Flat roofs

The same main subdivision of coverings and construction can be made as before. Example 8 shows the measurement of asphalt flat roofs. The measurement of metal covered roofs should not prove too difficult provided that their construction is fully understood.

Metal roofs are measured their net visible area with extra over linear items measured for drips, welts, rolls, seams, laps, and upstands. No deductions are made for voids not exceeding $1\,m^2$ within the area of the roof. When the opening has not been deducted, no further work around the opening has to be measured except in the case of holes. The pitch of the roof has to be stated in the description.

Flashings are measured linear, stating the girth; dressing into grooves, ends, angles, and intersections is deemed included. Gutters are measured in a similar manner and labours are deemed to be included. Outlets are enumerated, stating their size, and again labours are deemed included.

Mastic asphalt roofing (Work Section 19) is measured the area in contact with the base and again no deductions are made for voids within the area which do not exceed $1\,m^2$. The pitch has to be stated in the description if sloping. The width of work has to be classified as $\leq 500\,mm$ wide and measured in linear metres or as $> 500\,mm$ wide and measured in square metres. Skirtings, fascias, aprons, channel and gutter linings, and coverings to kerbs are measured linear stating the exact girth on face. All labours to these items are deemed included.

Bituminous felt roof coverings (Work Section 17) are measured in square metres if $> 500\,mm$ wide, stating the pitch. Coverings $\leq 500\,mm$ wide are measured in linear metres.

Timber flat roof construction is measured in a similar manner to that of floors. The fall in the roof may be created by the use of firrings, which are measured linear, stating their nominal size. Gutter boards and sides are measured in linear metres if not exceeding 600 mm wide, giving their thickness in the description. Over 600 mm wide, they are measured in square metres and the thickness stated. Timber drips and rolls, if required for joints in metal coverings, are measured linear, stating their size. Fascias, soffits, and rainwater goods are taken in a similar manner to those for pitched roofs.

Concrete roofs are measured in a similar manner to floors as described in Chapter 14. The fall in the roof is usually created in the screed; however, if the slab is sloping, then this is classified as either exceeding or not exceeding $15°$. If the slab exceeds $15°$, formwork has to be measured to the upper surface. Concrete upstands are measured as cubic items; formwork to the sides is measured in square metres. Trowelling the surface of the concrete is measured in square metres.

## Example 6   Trussed Rafter Roof

*Approach to measurement*

This roof comprises trussed rafters spaced between gables walls at each end. The number of trusses is calculated in the same way as the joists explained in Chapter 10. The roof coverings are measured the full length of the ridge, taken over the gable walls and timesed by the roof slope, then timesed by two for both slopes (Figure 29).

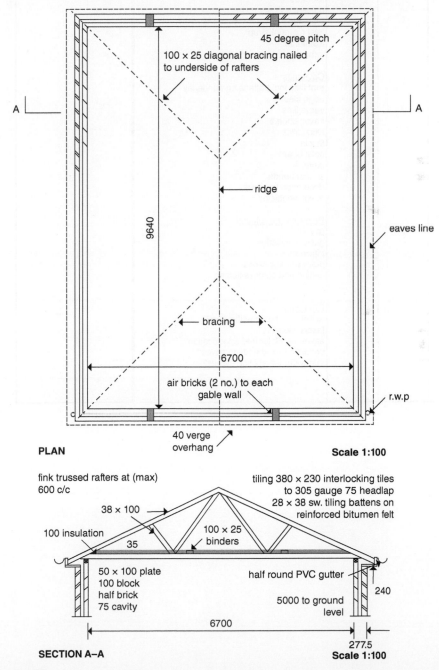

**PLAN**                                    **Scale 1:100**

**SECTION A–A**                             **Scale 1:100**

**Fig. 29**

Example 6

Pitched roof take-off

Taking-off list

Structure
wallplates
roof trusses
binders
bracing
joist restraint strap
insulation
airbrick

Coverings
roof tiles inc. battens and underlay
ridge tiles
verge tiles
eaves course
tilting fillet
fascia
soffit board
battens
spandril ends
decoration to eaves
eaves ventilator

Rainwater installation
gutters
gutter ancillaries
pipework
pipework ancillaries
testing and commissioning

RFI notes
wallplate strap specification?
fascia width?
eaves ventilator and specification?
detail for boxed ends to eaves?
rain water installation details?

| | | | | |
|---|---|---|---|---|
| 2/ | 9.64 | 19.28 | Primary or structural timbers, nominal size 100 × 75 mm, wallplate 16.1.1.3.0 | |
| 2/ | 5 | 10 | Metal fixings, wallplate strap, fixed to masonry 16.6.1..4.2 | |

|  | |
|---:|---:|
|  | 9.640 |
| divide by centres say | 2.000 |
|  | 4.820 |
|  | 5 |

| | | | |
|---|---|---|---|
| 17 | 17 | Engineered or prefabricated items, as figure 29, roof trusses 16.2.1.1.0 | |

|  | <u>lgth</u> |
|---:|---:|
|  | 9.640 |
| <u>ddt</u> air gap b.s. 2/0.05 | 0.100 |
|  | 9.540 |
| <u>ddt</u> half rafter width b.s. 2/0.5/0.038 | 0.038 |
|  | 9.502 |
| divide by spacing | 0.600 |
|  | 15.837 |
|  | 16 |
| add to convert from spaces to quantity | 1 |
|  | 17 |

Backing and other first fix timbers, nominal size 100 × 25 mm, diagonal bracing 16.3.1.2.0

| | | | |
|---|---|---|---|
| 2/ 2/ | 9.64 | 19.28 | (horizontal binders |
| 2/ | 5.44 | 21.78 | (sloping bracing |
| | | 41.06 | |

|  | <u>dia. bracing</u> |  |
|---:|---:|---:|
| half span | | |
| 0.5/6.700 | 3.350 | |
| add wallplate | 0.100 | |
|  | 3.450 | |

|  | <u>roof slope</u> |  |
|---:|---:|---|
|  | 4.212 | based on half span divided by cos 35 degrees |

|  | <u>lgth</u> |  |
|---:|---:|---|
|  | 5.444 | based on pythagoras using the two dimensions above |

*(Continued)*

(*Continued*)

| | | | | |
|---|---|---|---|---|
| | 8.99<br>6.70 | 60.26 | Quilt insulation, thickness 100 mm, laid between joists at 600 mm centres, horizontal<br>31.3.1.3.1 | |
| | | | ab       9.640<br>ddt roof trusses<br>17/0.038    0.646<br>            8.994 | area of insulation excludes area of joists forming part of roof truss |
| | 4 | 4 | Proprietary and individual spot items, airbrick, 225 × 150 mm, built into cavity wall<br>14.25.1.1.1 | |
| 2/ | 10.28<br>4.78 | 98.28 | Roof coverings, pitch 35 degrees, including battens and felt<br>18.1.1.1.0 | |
| | | | <u>lgth</u><br>           9.640<br>cavity wall b.s.<br>2/0.278    0.555<br>        10.195<br>verge b.s.<br>2/0.04    0.080<br>        10.275 | |
| | | | <u>slope lgth</u><br>span     6.700<br>cavity wall b.s. ab   0.555<br>        7.255<br>eaves b.s.<br>2/0.240    0.480<br>        7.735<br>overhang into gutter b.s.<br>2/0.050    0.100<br>0.5/    7.835<br>0.5 span    3.918 | |
| | | | <u>roof covering slope length</u><br>           4.782 | based on half span divided by cos 35 degrees |
| 2/ | 10.28 | 20.55 | Boundary work, eaves, horizontal<br>18.3.1.2.1 | |
| 2/ 2/ | 4.78 | 19.13 | Boundary work, verge, sloping<br>18.3.1.4.2 | |

| | | | | |
|---|---|---|---|---|
| | 10.28 | 10.28 | Boundary work, ridges, horizontal 18.3.1.3.1 | |

| | | | | |
|---|---|---|---|---|
| 2/ | 10.28 | 20.55 | Fascias to eaves, not exceeding 600 mm wide, finished width XXX, thickness 25 mm, horizontal, chamfered and grooved 16.4.1.1.1 | Width not shown on detail so 'XXX' used until RFI query answered. |

&

Soffits to eaves, not exceeding 600 mm wide, finished width 227 mm, horizontal 16.4.1.1.1

| | |
|---|---|
| soffit width | 0.240 |
| ddt fascia | 0.025 |
| | 0.215 |
| add tongue say | 0.012 |
| | 0.227 |

| | | | |
|---|---|---|---|
| 2/ | 22 | 44 | Eaves ventilator, Redland, nailed to rafter 16.6.2.8.1 |

| | |
|---|---|
| eaves vent qty | 10.195 |
| vent width say | 0.485 |
| | 21.021 |
| | 22 |

| | | | |
|---|---|---|---|
| 2/ | 10.20 | 20.39 | Backing and other first fix timbers, 38 × 50 mm, fixed to masonry 16.3.1.2.0 |

| | | | |
|---|---|---|---|
| | 4 | 4 | Ornamental ends, boxed end to eaves 16.5.1.0.0 |

| | | | | |
|---|---|---|---|---|
| 2/ | 10.28 | | Painting general surfaces, | Assumed no requirement |
| | 0.41 | 8.43 | > 300 mm girth, external | to be painted before fixing. |
| 4/ | 0.24 | | 29.1.2.2.0 | |
| | 0.15 | 0.14 | (box ends to eaves | |
| | | 8.57 | | |

| | |
|---|---|
| girth | |
| fascia width say | 0.150 |
| soffit | 0.240 |
| downstand say | 0.020 |
| | 0.410 |

*(Continued)*

| | | | Example 7 | |
|---|---|---|---|---|
| | | | | |
| | | | Structure | |
| | | | wallplates | |
| | | | rafters | |
| | | | joists | |
| | | | ridge | |
| | | | hip | |
| | | | valley | |
| | | | | |
| | | | Coverings | |
| | | | roof tiles inc. battens and underlay | |
| | | | ridge tiles | |
| | | | hip tiles | |
| | | | valley tiles | |
| | | | verge tiles | |
| | | | eaves course | |
| | | | tilting fillet | |
| | | | fascia | |
| | | | soffit board | |
| | | | battens | |
| | | | spandril/boxed ends | |
| | | | decoration to eaves | |
| | | | eaves ventilator | |
| | | | insulation | |
| | | | | |
| | | | Rainwater installation | |
| | | | gutters | |
| | | | gutter ancillaries | |
| | | | pipework | |
| | | | pipework ancillaries | |
| | | | testing and commissioning | |
| | | | | |
| | | | RFI Notes | |
| | | | airgap for rafter? | |
| | | | joist length? | |
| | | | ridge tile detail? | |
| | | | ridge/hip tile saddle detail? | |
| | | | all rf boundary work details? | |
| | | | tilting fillet detail? | |
| | | | hip iron detail? | |
| | | | eaves details? | |
| | | | boxed ends to eaves? | |
| | | | wallplate straps? | |
| | | | rainwater installation details? | |

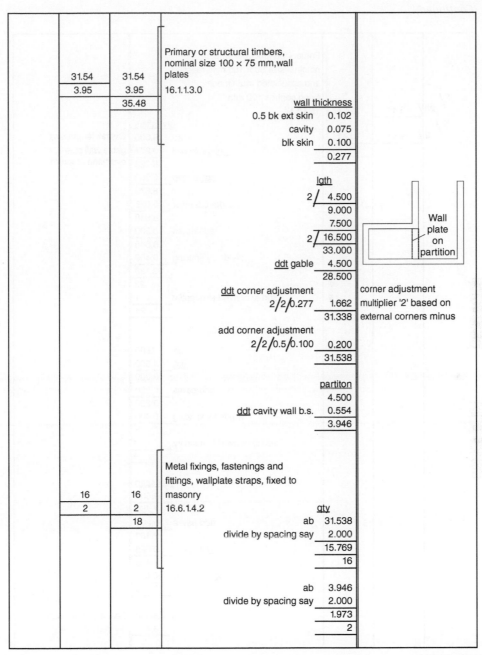

| | | | |
|---|---|---|---|
| 31.54 | 31.54 | Primary or structural timbers, nominal size 100 × 75 mm,wall plates | |
| 3.95 | 3.95 | 16.1.1.3.0 | |
| | 35.48 | | |

wall thickness

| | |
|---|---|
| 0.5 bk ext skin | 0.102 |
| cavity | 0.075 |
| blk skin | 0.100 |
| | 0.277 |

lgth

| | | |
|---|---|---|
| 2/ | 4.500 | |
| | 9.000 | |
| | 7.500 | |
| 2/ | 16.500 | |
| | 33.000 | |
| ddt gable | 4.500 | |
| | 28.500 | |

ddt corner adjustment

| | | corner adjustment |
|---|---|---|
| 2/2/0.277 | 1.662 | multiplier '2' based on |
| | 31.338 | external corners minus |

add corner adjustment

| | |
|---|---|
| 2/2/0.5/0.100 | 0.200 |
| | 31.538 |

partiton

| | |
|---|---|
| | 4.500 |
| ddt cavity wall b.s. | 0.554 |
| | 3.946 |

| | | | |
|---|---|---|---|
| 16 | 16 | Metal fixings, fastenings and fittings, wallplate straps, fixed to masonry | |
| 2 | 2 | 16.6.1.4.2 | |
| | 18 | | |

qty

| | |
|---|---|
| ab | 31.538 |
| divide by spacing say | 2.000 |
| | 15.769 |
| | 16 |

| | |
|---|---|
| ab | 3.946 |
| divide by spacing say | 2.000 |
| | 1.973 |
| | 2 |

Wall plate on partition

*(Continued)*

(*Continued*)

| | | | | | |
|---|---|---|---|---|---|
| | | | Primary or structural timbers, nominal size 50 × 150 mm, rafters and associated roof timbers, fixing centres 400 mm 16.1.1.1.4 | | |
| 2/ 24/ | 3.41 | 163.77 | | | |
| 2/ 9/ | 3.41 | 61.41 | | | |
| | | 225.18 | | | |

| | |
|---|---|
| qty rafters | |
| ab 9.000 | Overall length less |
| ddt gable wall 0.277 | gable wall plus |
| 8.723 | overhang at return. |
| ddt air gap 0.050 | |
| 8.673 | |
| ddt 0.5 rafter 0.025 | |
| 8.648 | |
| add eaves 0.200 | |
| 8.848 | |
| divide by spacing 0.400 | |
| 22.120 | |
| 23 | |
| add to convert to quantity 1 | |
| 24 | |
| | |
| ab 7.500 | |
| ddt 4.500 | |
| 3.000 | |
| add eaves 0.200 | |
| 3.200 | |
| divide by spacing 0.400 | |
| 8 | |
| add to convert to quantity 1 | |
| 9 | |
| | |
| half span | |
| 0.5/ 4.500 | |
| 2.250 | |
| add eaves 0.200 | |
| 2.450 | |
| ddt fascia 0.025 | |
| 2.425 | |
| ddt 0.5 ridge 0.013 | |
| 2.413 | |
| | |
| roof slope | Based on half span |
| 3.412 | divided by cos 45°. |

| | | | |
|---|---|---|---|
| 6.57 | 6.57 | Primary or structural timbers, nominal size 25 × 225 mm, rafters and associated timbers, ridge board | |
| 3.00 | 3.00 | 16.1.1.1.0 | |
| | 9.57 | | |

|  | ridge lgth |
|---|---|
| | 4.500 |
| add half span | 2.250 |
| | 6.750 |
| ddt bk & cavity | 0.177 |
| | 6.573 |

|  |  |
|---|---|
| | 7.500 |
| ddt half span twice | 4.500 |
| | 3.000 |

| | | | |
|---|---|---|---|
| 3 / 4.18 | 12.54 | Primary or structural timbers, nominal size 25 × 225 mm, rafters and associated timbers, hip and valley rafters 16.1.1.1.0 (hips | |
| 4.18 | 4.18 | (valley | |
| | 16.71 | | |

| | |
|---|---|
| slope lgth ab | 3.412 |
| half span ab | 2.413 |
| hip/valley lgth | |
| | 4.179 |

Based on pythagoras

3486 (rafter length) · Hip · 2412 ½

(*Continued*)

*(Continued)*

| | | | |
|---|---|---|---|
| | | | Primary or structural timbers, nominal size 50 × 125 mm, roof and floor joists, fixing centres 400 mm |
| 22/ | 4.50 | 97.83 | 16.1.1.4.4 |
| 8/ | 4.50 | 36.00 | |

|  | qty jsts |
|---|---|
| ab | 9.000 |
| <u>ddt</u> cavity wall b.s. | |
| 2/0.277 | 0.554 |
| | 8.446 |
| <u>ddt</u> air gap b.s. | |
| 2/0.050 | 0.100 |
| | 8.346 |
| <u>ddt</u> 0.5 joist b.s. | 0.050 |
| | 8.296 |
| divide by spacing | 0.400 |
| | 21 |
| add 1 to convert to quantity | 1 |
| | 22 |
| | |
| | 7.500 |
| | 4.500 |
| | 3.000 |
| <u>ddt</u> blk ptn | 0.100 |
| | 2.900 |
| <u>ddt</u> air gap b.s. | |
| 2/0.050 | 0.100 |
| | 2.800 |
| divide by spacing | 0.400 |
| | 7 |
| add 1 to convert to quantity | 1 |
| | 8 |

133.83

| | | | |
|---|---|---|---|
| 2/ | 12.33 | | Roof coverings, pitch 45 degrees, |
| | 3.54 | 87.15 | inlcuding underlay and battens |

18.1.1.1.0

| | 2/ | 4.500 |
|---|---|---|
| | | 9.000 |
| | | 7.500 |
| | | 16.500 |

<u>ddt</u> corner adjustment

| 1/2/0.5/4.500 | 4.500 |
|---|---|
| | 12.000 |
| add verge o/hang | 0.075 |
| | 12.075 |
| add eaves | 0.200 |
| | 12.275 |
| add o/hang into gutter | 0.050 |
| | 12.325 |

<u>covering slope lgth</u>

| <u>half span</u> | |
|---|---|
| | 4.500 |
| add eaves b.s. | |
| 2/0.200 | 0.400 |
| | 4.900 |
| add o/hang into gutter b.s. | |
| 2/0.05 | 0.100 |
| 0.5/ | 5.000 |
| | 2.500 |

| <u>slope lgth</u> | Based on half span |
|---|---|
| 3.536 | divided by cos 45°. |

| | | | |
|---|---|---|---|
| | 9.83 | 9.83 | Boundary work, ridge, horizontal |

18.3.1.3.1

| | <u>lgth</u> |
|---|---|
| ab | 6.750 |
| ab | 3.000 |
| | 9.750 |
| add verge overhang | 0.075 |
| | 9.825 |

*(Continued)*

Rafter · 45° · Clg. Jst. · Plate · 200

(*Continued*)

| | | | | |
|---|---|---|---|---|
| 3/ 4.33 | 12.99 | Boundary work, hips, sloping<br>18.3.1.6.2 | | |
| | | covering slope ab | 3.536 | |
| | | covering half span ab | 2.500 | |
| | | <u>hip/valley lgth</u> | | |
| | | | 4.330 | Based on pythagoras |
| 4.33 | 4.33 | Boundary work, valley, sloping<br>18.3.1.5.2 | | |
| 2/ 3.54 | 7.07 | Boundary work, verges, sloping<br>18.3.1.6.2 | | |
| 29.65 | 29.65 | Boundary work, eaves, horizontal<br>18.3.1.2.1 | | |
| | | ab | 28.500 | |
| | | add corner adjustment | | Based on eaves plus |
| | | 2/2/0.250 | <u>1.000</u> | overhang into gutter. |
| | | | 29.500 | |
| | | add for verge b.s. | <u>0.150</u> | |
| | | | 29.650 | |
| 3 | 3 | Fittings, hip irons<br>18.4.1.4.0 | | |
| 29.45 | 29.45 | Backing and other first fix<br>timbers, tilting fillet<br>16.3.1.2.0 | | |
| | | ab | 29.650 | |
| | | <u>ddt</u> corner adjustment | | |
| | | 2/2/0.05 | <u>0.200</u> | |
| | | | 29.450 | |

| | | | |
|---|---|---|---|
| | 29.40 | 29.40 | Fascia, not exceeding 600 mm wide, finished width 175 mm, thickness 25 mm, horizontal<br>16.4.1.1.0 |

ab 29.450
ddt corner adjustment
2/2/0.5/0.025   0.050
_____
29.400

| | | | |
|---|---|---|---|
| | 29.00 | 29.00 | Soffit, not exceeding 600 mm wide, finished width 175 mm, thickness 19 mm, horizontal<br>16.4.1.1.0 |

ab 29.400
ddt corner adjustment ab   0.050
_____
29.350
ddt corner adjustment
2/2/0.5/0.175   0.350
_____
29.000

| | | | |
|---|---|---|---|
| | 28.50 | 28.50 | Backing and other first fix timbers, nominal dimension 38 × 50 mm, battens<br>16.3.1.2.0 |

| | | | |
|---|---|---|---|
| 2/ | 0.20<br>0.20 | 0.08 | Boarding, not exceeding 600 mm wide, finished width 175 mm, thickness 25 mm, vertical<br>16.4.1.3.0<br>(boxed ends to eaves |

Painting general surfaces, > 300 mm girth, external
29.1.2.2.0

| | | | |
|---|---|---|---|
| | 29.40<br>0.38 | 11.03 | |
| 2/ | 0.20<br>0.20 | 0.08 | |
| | | 11.11 | |

girth
fascia 0.175
soffit 0.200
_____
0.375

*(Continued)*

*(Continued)*

| | | | |
|---|---|---|---|
| | 11.45 | | Insulation quilt, thickness 150 mm, laid between joists at 400 mm centres, horizontal |
| | 3.95 | 45.17 | 31.3.1.3.1 |
| <u>ddt</u> | | | |
| 30/ | 0.05 | | (cgl jsts |
| | 3.95 | 5.87 | |
| | | 39.30 | |

ab    12.000
<u>ddt</u> cavity wall b.s.
2/0.277    0.554
        11.446

        4.500
<u>ddt</u> cavity wall b.s.
2/0.277    0.554
        3.946

| | | | |
|---|---|---|---|
| | | | <u>Rainwater installation</u> |
| | 29.65 | 29.65 | Gutters, nominal size 100 mm, straight, fixed to timber 33.5.1.1.1 |
| | 2 | 2 | Gutter ancillaries, stop ends 33.6.1.0.0 |
| | 4 | 4 | Gutter ancillaries, angles 33.6.1.0.0 |
| | 4 | 4 | Gutter ancillaries, outlets 33.6.1.0.0 |
| 4/ | 6.00 | 24.00 | Pipework, nominal diameter 68 mm, straight, fixed to masonry 33.1.1.1.1 |

| | | | |
|---|---|---|---|
| | 4 | 4 | Pipework ancillaries, offsets<br>33.2.1.0.0 |
| | 4 | 4 | Pipework ancillaries, shoes<br>33.2.1.0.0 |
| | 4 | 4 | Pipework ancillaries, balloon grating<br>33.2.1.0.0 |
| | Item | | Testing and commissioning, rainwater installation<br>33.10.1.0.0 |

(*Continued*)

| | | | | |
|---|---|---|---|---|
| | 4.26 | 4.26 | Skirtings, net girth on face XXX, vertical | |
| 2 / | 2.43 | 4.86 | 19.3.1.3.0 | |
| | | 9.12 | | |
| | 4.26 | 4.26 | Other boundary work, net girth on face XXX, vertical 19.6.1.3.0 | |
| | 4.26 | 4.26 | Edge trim, dimensioned description XXX, aluminium flashing, fixed to concrete | |

# 12

# Internal Finishes

## Schedules

A schedule of internal finishes listing floor by floor and room by room the ceiling, wall, and floor finishes and decorations is an invaluable aid to the measurement of this section of the work. Details of cornices, skirtings, dados, and the like can be added to the schedule as required. The schedule brings to light any missing information, enables rooms with similar finishes to be grouped together more easily, and reduces the need to refer to the specification during the taking-off. Looping through the entries on the schedule as the items are measured ensures that none are missed. At a later stage, the schedule provides a useful quick reference indicating what has been taken without one having to look through the dimensions. An example of a typical schedule is given in Figure 32. This is an information schedule and may well be provided by the architect. Another schedule could be utilised to measure the finishes. Here, girths and heights of rooms would be recorded and summarised to reduce any repetitive dimensions.

Refer to Example 9 for how to approach the measurement of simple finishes without the use of a schedule.

| Internal finishes schedule | | | | | |
|---|---|---|---|---|---|
| | Ceiling | | Walls | | |
| Location | Finish | Decoration | Finish | Decoration | Dado |
| Bathroom<br><br>Kitchen etc. | 9.5 mm plasterboard 5 mm skim | 2 ct emulsion | 13 mm two ct plaster | 2 ct emulsion | Glazed tiling 1 m high |
| | | | Floor | | |
| Location | Cornice | Skirting | Bed | Finish | Remarks |
| Bathroom<br><br>Kitchen etc. | 250 mm gth moulded | 25 × 100 mm rounded s/w Kps & 3 | 35 mm cement and sand | Carpet | Area n.e. 4 m² |

Fig. 32

*Willis's Elements of Quantity Surveying*, Fourteenth Edition. Roy Hills and Sandra Lee.
© 2024 John Wiley & Sons Ltd. Published 2024 by John Wiley & Sons Ltd.

## Subdivision

If the ceiling heights and construction vary throughout the building, it may be advisable to keep the measurements on each floor separate. If, however, the room layout and finishes repeat from floor to floor, it is probably more expedient to use timesing. As this will quite possibly be the largest section of taking-off, care should be taken to signpost the dimensions to enable measurements to be traced easily at a later stage. The principal subdivisions in the order recommended for measuring are:

- Floor finishes and screeds (unless taken with floor construction).
- Ceiling finishes (including attached beams with the same finish).
- Isolated beams.
- Wall finishes (including attached columns with the same finish).
- Isolated columns.
- Cornices.
- Dados.
- Skirtings.

*Note*: It may be convenient to measure floors with ceilings if the areas are the same. Confusion may arise, however, if the specification changes are not consistent between the two, and also if deductions have to be made from floors only for such items as stair openings, hearths, and fittings.

## Generally

The measurements for in situ finishes are taken as those of the base to which the finish is applied. Normally, the structural room sizes are taken for the measurement of finishes and decoration. Generally, no deduction is made for voids within the area of the work when they do not exceed 1.0 m².

## Floor finishes

In situ sheet and tile finishings all have to be described as follows:

- Level and to falls only ≤ 15° from horizontal.
- To falls, cross falls, and slopes ≤ 15° from horizontal.
- To falls, cross falls, and slopes > 15° from horizontal.

If the work is laid in bays, this has to be stated. There is no longer a need to keep work in staircase areas and plant rooms separate.

Skirtings are usually measured with internal finishes, as their length relates to that of the walls. Floor coverings in door openings are taken either with doors, as the surveyor dealing with these is familiar with the dimensions or with the finishes. It should be remembered that dividing strips may be required where the finish changes. Doors are usually hung flush with the face of the wall of the room into which they open; therefore, the floor finish in the opening should be that of the room that shows when the door is closed. Mat frames and mat wells should be measured with floor finishes.

Screeds, if required, are measured with floor finishes as their areas are usually the same. Care should be taken to ascertain the thickness of screeds where the floor finishes vary. Usually, it is necessary to have floor surfaces level; differing finishes may have varying thicknesses, requiring the screed to take up the difference. Sometimes this difference is allowed for in the floor construction.

Timber boarding, plywood, or chipboard to floors is measured in square metres unless it does not exceed 600 mm wide when it is measured linear. No deduction is made for voids not exceeding $1.0\,m^2$ within the area of the flooring. Remember that nosings and margins may be required around openings in the floor.

Insulation may also be required and it is also measured using the rules in this Work Section.

## Ceiling finishes

Ceiling areas are taken from wall to wall in each room using the figured structural dimensions. The dimensions are usually best taken over the beams, and the adjustments made later. Care should be taken as the work to the soffit of beams may well be in a different height classification from the main ceiling of a room. If the majority of ceiling finishings are identical, it may be advantageous to measure the ceiling over all the internal walls and partitions. From this measurement will be deducted the total lengths of the internal walls and partitions, possibly obtained from previous dimensions, squared by their respective thicknesses. Treating the surface of concrete by mechanical means, or as it is traditionally known *hacking concrete* to receive plaster, is measured superficially and is conveniently *anded on* to the finishes dimensions. The application of bonding agents or other preparatory work is described with the plasterwork. Plasterboard linings to ceilings are measured in square metres.

## Wall finishes

In the case of wall finishings, it is best to schedule or total in waste calculations the perimeters of all rooms having the same height and finish. The areas of the walls may thus be contained in two or three dimensions instead of the large number there would be if each room were measured separately. As in the case of the measurement of walls, openings are generally ignored, the wall finish being measured across. Deductions for openings together with work to reveals would be taken later with the measurement of windows, doors, and blank openings. If large openings or infill panels occur from floor to ceiling height, it is sometimes preferable to measure the finishes net. Care must be taken to inform other measurers which openings have been measured net so as to avoid the possibility of a double deduction. As a general rule, if the walling has been measured across an opening, so should the finishes be.

If in situ wall finishes are applied to brick or block walls, then the raking out of the joints to form a key is deemed included. Preparation of concrete walls is dealt with as mentioned for ceilings in this chapter.

## Angle beads and so on

All work in forming returns, ends, and internal and external angles are deemed included. Designed movement beads and other beads such as angle beads are measured as linear items. Sizes are given in the description, but working finishes to beads are deemed included.

## Decoration

The decoration to ceilings and walls should be measured with the finish. If the decoration is all the same, this can be anded on to the plaster or plasterboard description. If, however, there is a mixture of several types of decoration, such as emulsion, gloss paint, and wallpaper, it will probably be best to subdivide the plaster or plasterboard measurement so that the appropriate decoration can be anded on. Where all is, say, emulsion except wallpaper in one or two rooms, it is usually more satisfactory to measure the whole as emulsion in the first instance and then to deduct the emulsion and add wallpaper for the appropriate area. Isolated paintwork where the girth does not exceed 300 mm or where the area does not exceed 1.0 m$^2$ is measured linear or enumerated, respectively. Paintwork to walls, ceilings, beams, and columns is described as work to general surfaces and, if the same, can be added together in the bill. Work to ceilings and beams exceeding 3.5 m above floor level has to be so described.

## Cornices and coves

The measurement of cornices and coves is usually straightforward, as they normally run all round the room without interruption. Both in situ and prefabricated types are measured the length in contact with the base, using the length already calculated for the perimeter of the walls. The measurement must include returns to projections and to beams if these occur in the ceiling. Ends, angles, and intersections are deemed included. Paintwork to cornices does not have to be kept separate if it is the same as that to the walls or ceilings. In fact, it is questionable whether any adjustment should be made to the wall and ceiling decoration if this is the same, particularly if the cornice is contoured.

## Skirtings

There is a difference of opinion as to whether skirtings should be measured net or measured gross across openings. The disadvantage with measuring net is that the taker-off measuring finishes will have to ascertain the width of openings and possibly the depth of reveals. On the other hand, wall decoration, as opposed to plaster, should be deducted behind the skirting. Thus, if the skirtings are measured gross with the deduction of decoration behind, then the doors are measured a deduction of plaster and decoration is usually made for the height of the opening. This results in the decoration being deducted twice behind the skirting, and therefore an addition of decoration will have to be made for the length by the height of the skirting across the door opening.

Timber skirtings are measured linear, usually using structural wall face dimensions, but care will have to be exercised with measurements to isolated piers and the like, as the length will be increased. The size and shape of the skirting and nature of background are given in the description but mitres, ends, and intersections are deemed included. If the skirting is fixed to grounds, no deduction of plaster is made for grounds. Dado and picture rails are measured in the same way as skirtings.

In-situ and tile skirtings are measured as linear items, stating the height and thickness in the description. Fair and rounded edges are deemed to be included, as are ends and angles.

## Wall tiling

First, it is necessary to ascertain whether the wall tiling is fixed with adhesive or bedded in mortar. The specification may, particularly in the case of a tiled dado, call for the wall behind the tiling to be plastered like the remainder of the wall and for the tiles to be fixed to the plaster with adhesive. Alternatively, there may be a cement and sand screed behind the tiling, with the tiles either bedded in mortar or fixed with adhesive. If there is a mixture of wall tiling and plaster finish on a wall, it is usually necessary for the back of the tiles to be flush with the face of the plaster. If special tiles with rounded edges are specified at top edges and external angles, these are enumerated and dimensioned descriptions given. The rules for measurement of tiling follow closely those for in situ finishes.

## Internal partitions

The measurement of timber stud partitions is described in Chapter 9. Metal stud partitions can be measured in a similar way.

## Dry wall linings

Plasterboard dry linings are measured in square metres, stating if the width on face is over or under 300 mm wide. Internal and external angles are measured as linear items.

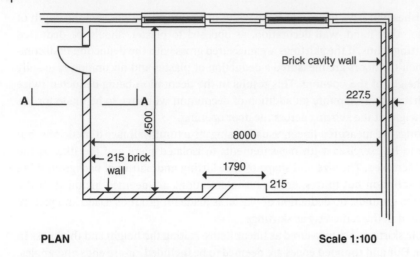

**PLAN**                                                              **Scale 1:100**

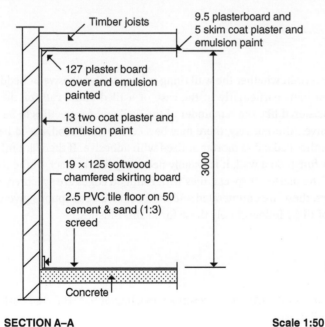

**SECTION A–A**                                                      **Scale 1:50**

**Fig. 33**

**Example 9    Internal Finishes (Figure 33)**

| | | | | |
|---|---|---|---|---|
| | | | Example 9 | |
| | | | Taking-off list<br>ceilings pla.bd & skim<br>paint to clgs<br>floor tiles<br>screed<br>wall plaster<br>paint to walls<br>plaster cove<br>skirting<br>decoration to skrtg | |
| | | | RFI notes<br>screed layers? | |
| | | | Finish to ceilings, plasterboard 9.5 mm and skim coat 5 mm, > 600 mm wide<br>28.9.2.0.0 | |
| | 8.00<br>4.50 | 36.00 | | |
| ddt | 1.79<br>0.22 | 0.38 | & | Pier still deducted despite being ≤ 1 m² because of location at perimeter. |
| | | 35.62 | Painting general surfaces, > 300 mm girth, internal<br>29.1.1.1.0 | |
| | | | & | |
| | | | Finish to floors, PVC tile 2.5 mm, > 600 mm wide, level and to falls only ≤ 15° from horizontal, screed background<br>28.2.2.1.1 | |
| | | | & | |
| | | | Screeds, beds, and toppings, thickness 50 mm in single layer, > 600 mm wide, level and to falls only ≤ 15° from horizontal, concrete background<br>28.1.1.1.2 | |

*(Continued)*

*(Continued)*

| | | | | |
|---|---|---|---|---|
| 2/ | 8.00 | | Finish to walls, two coat plaster 13 mm, > 600 mm wide, masonry background 28.7.2.0.0 | |
| | 3.00 | 48.00 | | |
| 2/ | 4.50 | | | |
| | 3.00 | 27.00 | & | |
| | | 75.00 | | |
| | | | Painting general surfaces, > 300 mm girth, internal 29.1.2.1.0 | |
| 2/ | 3.00 | 6.00 | Finish to walls, two coat plaster 13 mm, ≤ 600 mm wide, masonry background 28.7.1.0.0 | |
| | | | & | |
| | | | Beads, angle beads, masonry background 28.28.1.0.2 | |
| | | | & | |
| | | | Painting general surfaces, ≤ 300 mm girth, internal 29.1.1.1.0 | |
| 2/ | 8.00 | 16.00 | Precast plaster components, cove 127 mm girth, fixed using adhesive, plaster background 28.36.1.1.1 | Angles deemed included. |
| 2/ | 4.50 | 9.00 | | |
| 2/ | 0.22 | 0.43 | & | |
| | | 25.43 | | |
| | | | Skirting, 19×125 mm chamfered 22.1.1.0.0 | |
| | | | & | |
| | | | Painting general surfaces, ≤ 300 mm girth, internal 29.1.1.1.0 | |
| | | | <u>ddt</u> Painting general surfaces, > 300 mm girth, internal, abd | |
| | 25.43 | | | |
| | 0.13 | 3.18 | | Paint behind skirting not required. |

# 13

# Windows and Doors

## Subdivision

The measurement of windows and doors can be conveniently subdivided as follows:

- Windows
- External doors
- Internal doors
- Blank openings.

Whilst it is not essential to take external and internal doors separately, it will probably be found that they are different types of doors with different finishes, and therefore they may have little in common.

Windows

- Timber, metal or unplasticised polyvinyl chloride (uPVC) casement and fixing.
- Glass including double glazed units (if not included in the above).
- Ironmongery (if not included in the above).
- Decoration (if required).

Opening Adjustments

- Deduction of brickwork and blockwork and external and internal finishes.
- Support to the work above the window.
- Damp proofing and finishes to the head externally and internally.
- Damp proofing and finishes to the external and internal reveals.
- Damp proofing and finishes to the external and internal cills.

External Doors

- The same items would be measured for external doors as shown above for windows.

Internal Doors and Blank Openings

- Again, this will be similar as given for windows, but in place of cills, flooring in the opening will be required. Damp proofing around the opening will not be necessary.

*Willis's Elements of Quantity Surveying*, Fourteenth Edition. Roy Hills and Sandra Lee.
© 2024 John Wiley & Sons Ltd. Published 2024 by John Wiley & Sons Ltd.

Steps and threshold details to external doors may be measured at this stage along with porches, canopies, or other special features.

If these items are followed through systematically in each group, there will be less chance of items being missed.

## Schedules

As with internal finishes measurement and for similar reasons, it is usually prudent to use a schedule. If not provided by the designer, a schedule of windows and doors could be prepared before commencing measurement. Before preparing the schedule, the windows should be lettered or numbered on the floor plans and the elevations. If there are several floors to the building, then a system of numbering should be devised which enables windows on a particular floor to be located readily. For example, a letter could be allocated to each floor followed by a number for each window, the numbering starting at a particular point on each floor and proceeding in, say, a clockwise direction round the building. Whilst windows are usually scheduled floor by floor, there may be good reason to work by elevations or by window types or even by a combination of all three. The total number of windows entered on the schedule should be checked carefully with the total number shown on the drawings to ensure that none is missed. A further check must be made to ensure that the windows shown on the elevations tie up with those shown on the plans. Discrepancies may occur because clerestory windows or windows to mezzanine floors are sometimes shown on the elevations but not on the plans. Any differences found should be mentioned to the architect and the matter resolved before commencing measurement. The schedule should aim to set out the details of each window so that it is hardly necessary to refer to the drawings during measurement. Usually, it should be possible to measure together in one group all windows of one type irrespective of their size, the wall thickness, and so on.

A note should be made at the commencement of the measurement for each group of the numbers of windows being dealt with; care should be taken throughout that this total is accounted for in each item. A common fault of beginners is to separate the entire measurement of windows of the same type but of different sizes or in different thicknesses of walls. Such a method may give less trouble but will probably take longer. Although the grouping of windows of different sizes may require more concentration and care, proper use of the schedule will considerably simplify the work. Doors on the plans should be lettered or numbered socially as described for windows above. External doors may be included with the window numbering, and internal blank openings included with internal doors.

## Timesing

Window dimensions will probably contain a fair amount of timesing, and great care is necessary to ensure that this is done correctly. As each of the subdivisions of measurement is completed, it is advisable to total the timesing of each item to ensure that it equals the number of windows being measured. If, for instance, 20 windows are being measured in a group, timesing of deductions, lintels, cills, and the like should total 20 unless a change in specification or design for a particular window requires a smaller number.

## Special features

Usually, any special features that definitely relate to the window (such as small canopies above or decorative brickwork underneath) will be measured with the window.

## Dormer windows

In the case of dormer windows, the window itself may be taken with the other windows whilst the adjustment of the roof would normally be taken with the roofs. This division will generally be found to be convenient, particularly if the roofs and windows are being measured by different persons. The opening for a dormer window may sometimes be partly in the wall and partly in the roof, in which case the wall adjustment would be made with the window measurement, in the same way as for the other window openings.

## Adjustments

When measuring a group of items, the advantage of taking initially the same description for similar work and then making an adjustment for small differences often becomes evident in the measurement of windows.

For example, if all the windows are glazed in clear glass except for a small proportion which have patterned glass, the simplest way of measuring may be to take all as clear glass and then, checking over carefully with the schedule, make adjustment for those that need patterned glass. If, as is not impossible, the measurement of the patterned glass to one or more windows should be missed, then if clear glass has been measured to them all, the error will be much less than if none had been measured. Similarly, when making adjustments for the openings, different decoration may be applied to the walls. If the predominating one is chosen for the deductions to all windows, then, as an adjustment, the true decoration may be deducted where appropriate and the predominating finish, deducted earlier, added back. Apart from minimising the effect of errors, this method also facilitates the grouping of descriptions under the same measurements.

## Windows and doors

Timber, metal, and uPVC windows and frames or doors and frames are enumerated and described with a dimensioned diagram. A reference to a catalogue or standard specification may remove the need to provide a diagram. The method of fixing must be shown unless this is at the discretion of the contractor. Timber window boards and cover fillets are measured as linear items, stating their cross-section dimensions and labours in the description.

**Example 10   Window**

| | | | Example 10 | |
|---|---|---|---|---|
| | | | Taking-off list<br>window<br>window board<br>glazing<br>paint to wdw<br>paint to wdw bd<br>adjust cavity wall<br>lintel<br>cavity tray<br>brick flat arch<br>adjust bk wall for arch<br>close cavity<br>dpc to jambs<br>angle beads<br>plaster to reveals<br>paint to reveals<br>cill<br>adjust bk wall for cill<br>dpc to cill | |
| | | | RFI notes<br>dble glazed unit (dgu) glass type?<br>full wdw frame dimensions?<br>pcc lintel dims & reinf?<br>ss lintel spec & dims?<br>bk cill dims? | |
| | 1 | 1 | Windows and window frames,<br>1.77 × 1.35 m as Fig. 34, built into<br>brickwork<br>23.1.1.0.1 | |

| | | | |
|---|---|---|---|
| 1.77 | 1.77 | Window board, 25 × 150 mm, rebated and rounded on one edge<br>22.5.1.0.0 | |
| 2 | 2 | Sealed, double glazed units, clear glass, 503 × 1210 mm, 20 mm air gap<br>27.2.2.2.0 | Usually the window will be supplied with the dgu, but for this example, it is assumed that the dgu is supplied separately. |

width
1.770
<u>ddt</u> frame say
2/0.040  0.080
1.690
<u>ddt</u> mullions say
2/0.030  0.060
1.630
divided by number of panels  3
0.543
<u>ddt</u> side rails say
2/0.020  0.040
0.503

height
1.350
<u>ddt</u> head say  0.040
1.310
<u>ddt</u> cill say  0.050
1.260
<u>ddt</u> head rail say  0.020
1.240
<u>ddt</u> cill rail say  0.030
1.210

| | | | |
|---|---|---|---|
| 1 | 1 | Sealed, double glazed units, clear glass, 543 × 1240 mm, 20 mm air gap<br>27.2.2.2.0 | |

*(Continued)*

(*Continued*)

| | | | | |
|---|---|---|---|---|
| 1.77<br>1.35 | 2.39 | Painting glazed surfaces, ><br>300 mm girth, internal<br>29.2.2.1.0 | Assumed already primed. |
| | | & | |
| | | Painting glazed surfaces, ><br>300 mm girth, external<br>29.2.2.1.0 | |
| 1 | 1 | Painting general surfaces,<br>isolated areas ≤ 1 m² irrespective<br>of location or girth, internal<br>29.2.3.1.0 | |
| | | <u>window board</u><br>area =    0.266 | Measured as isolated<br>a readue to size ≤ 1 m². |
| 1.77<br>1.35 | 2.39 | <u>ddt</u><br>Walls, 102 mm thick, brick work<br>abd | abd = as before described |
| | | & | |
| | | <u>ddt</u><br>Forming cavity, abd | |
| | | & | |
| | | <u>ddt</u><br>Walls, 100 mm thick, block work abd | Includes insulation if<br>measured as a<br>composite item before. |
| | | & | |
| | | <u>ddt</u><br>Finish to walls, plaster, 13 mm<br>thick, > 600 mm wide, abd | |
| | | & | |
| | | <u>ddt</u><br>Painting general surfaces, ><br>300 mm girth abd | |

| | | |
|---|---|---|
| 1 | 1 | Precast concrete goods, lintel, 2.07 × 0.10 × 0.21 m 13.1.1.0.0 |

lintel lgth

| | |
|---|---|
| opg | 1.770 |
| add bearing b.s. | |
| 2/0.150 | 0.300 |
| | 2.070 |

cross-sectional area

| | | |
|---|---|---|
| | 0.021 | Lintel cross sectional area ≤ 0.50 m² so area of blockwork displaced not deducted. |

| | | |
|---|---|---|
| 2.07 | 2.07 | Isolated metal members, overall girth 0.375 m as fig. 34 26.1.1.1.0 |

ss lintel girth

| | |
|---|---|
| wedged into block say | 0.025 |
| face of block say | 0.150 |
| across cavity say | 0.100 |
| under brick skin say | 0.100 |
| | 0.375 |

| | | |
|---|---|---|
| 1.77 | 1.77 | Bands, soldier course, 0.225 m wide on face, flush, horizontal, entirely of stretchers 14.7.1.3.1 |

ddt
Walls, 102 mm thick, brickwork abd

| | | |
|---|---|---|
| 1.77 | | Area of facing brick |
| 0.23 | 0.40 | displaced by addition of soldier course. |

ddt
Finish to walls, plaster, overall thickness 13 mm, > 600 mm wide, masonry background, abd

| | |
|---|---|
| 2.07 | |
| 0.23 | 0.47 |

(*Continued*)

*(Continued)*

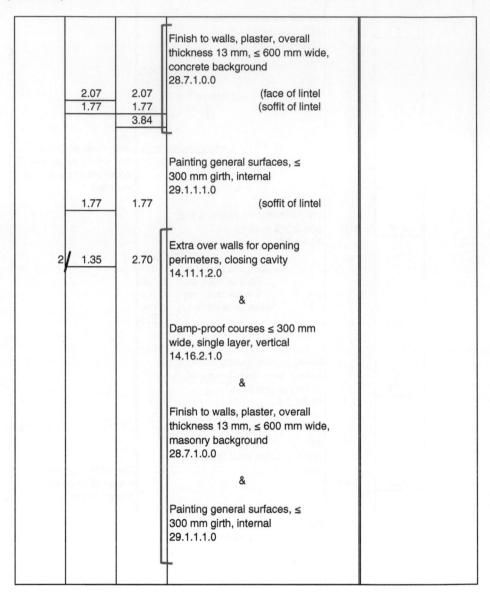

| | | | |
|---|---|---|---|
| | | | Finish to walls, plaster, overall thickness 13 mm, ≤ 600 mm wide, concrete background 28.7.1.0.0 |
| | 2.07 | 2.07 | (face of lintel |
| | 1.77 | 1.77 | (soffit of lintel |
| | | 3.84 | |
| | | | Painting general surfaces, ≤ 300 mm girth, internal 29.1.1.1.0 |
| | 1.77 | 1.77 | (soffit of lintel |
| 2/ | 1.35 | 2.70 | Extra over walls for opening perimeters, closing cavity 14.11.1.2.0 |
| | | | & |
| | | | Damp-proof courses ≤ 300 mm wide, single layer, vertical 14.16.2.1.0 |
| | | | & |
| | | | Finish to walls, plaster, overall thickness 13 mm, ≤ 600 mm wide, masonry background 28.7.1.0.0 |
| | | | & |
| | | | Painting general surfaces, ≤ 300 mm girth, internal 29.1.1.1.0 |

| | | | |
|---|---|---|---|
| | 1.77 | 1.77 | Beads, angle beads |
| 2/ | 1.35 | 2.70 | 28.28.1.0.0 |
| | | 4.47 | |
| | | | |
| | | | Bands, special brick cill, |
| | 1.77 | 1.77 | projecting, horizontal |
| | | | 14.7.3.3.0 |
| | | | |
| | | | <u>ddt</u> |
| | 1.77 | | Walls,102 mm thick, brick work abd |
| | 0.15 | 0.02 | |
| | | | |
| | | | Extra over walls for opening |
| | 1.77 | 1.77 | perimeters, closing cavity |
| | | | 14.11.1.2.0 |
| | | | |
| | | | Damp-proof courses ≤ 300 mm |
| | | | wide, single layer, horizontal, |
| | 1.77 | 1.77 | stepped |
| | | | 14.16.2.1.0 |

<u>girth</u>

vertical, rear face of cill say   0.150

under side of cill say   0.100

0.250

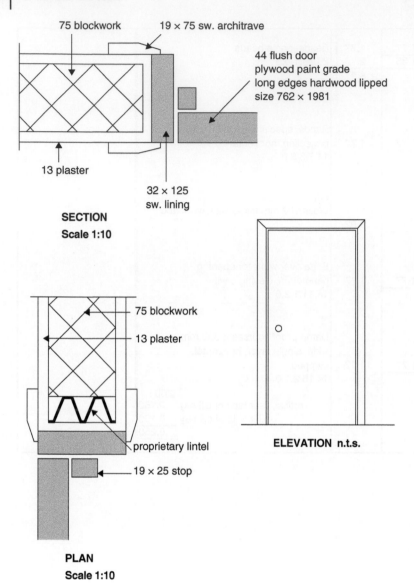

75 blockwork

19 × 75 sw. architrave

44 flush door
plywood paint grade
long edges hardwood lipped
size 762 × 1981

13 plaster

32 × 125
sw. lining

**SECTION**
**Scale 1:10**

75 blockwork

13 plaster

proprietary lintel

19 × 25 stop

**ELEVATION n.t.s.**

**PLAN**
**Scale 1:10**

**Fig. 35**

**Example 11    Internal Door**

| | | | | |
|---|---|---|---|---|
| | | | Example 11 | |
| | | | Taking-off list<br>dr unit<br>decoration<br>ironmongery<br>linings<br>stops & architrave<br>decoration<br>adjust opg, plaster & decs<br>lintel<br>adjust skrtg & decs | |
| | | | RFI notes<br>ironmongery specification? | |
| 1 | | 1 | Doors, 762 × 1981 × 40 mm, as fig. 35<br>24.2.1.0.0 | For this example, the assumption is made that the door is not supplied as a door set. |
| 2/ | 0.76<br>1.98 | 3.02 | Painting general surfaces, > 300 mm girth, internal<br>29.1.2.2.0 | |
| 2/ | 0.76<br>0.04 | 0.06 | | |
| 2/ | 1.98<br>0.04 | 0.16 | | |
| | | 3.24 | | |
| 1 | | 1 | Ironmongery, pair 75 mm pressed steel hinges, wood background<br>24.16.2.1.0 | |
| | | | & | |
| | | | Ironmongery, mortice lock<br>24.16.2.0.0 | |
| | | | & | |
| | | | Ironmongery, SAA lever furniture<br>24.16.2.0.0 | |

*(Continued)*

*(Continued)*

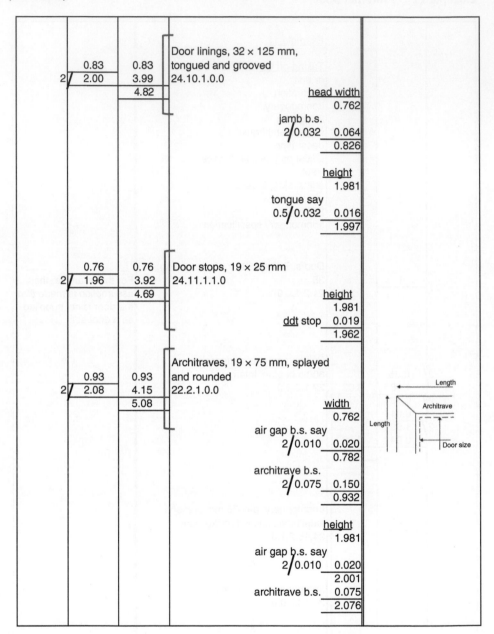

| | | | |
|---|---|---|---|
| | 0.83 | 0.83 | Door linings, 32 × 125 mm, |
| 2/ | 2.00 | 3.99 | tongued and grooved |
| | | 4.82 | 24.10.1.0.0 |

<u>head width</u>
0.762

jamb b.s.
2/0.032   0.064
          0.826

<u>height</u>
1.981

tongue say
0.5/0.032   0.016
            1.997

| | | | |
|---|---|---|---|
| | 0.76 | 0.76 | Door stops, 19 × 25 mm |
| 2/ | 1.96 | 3.92 | 24.11.1.1.0 |
| | | 4.69 | |

<u>height</u>
1.981

<u>ddt</u> stop   0.019
            1.962

| | | | |
|---|---|---|---|
| | 0.93 | 0.93 | Architraves, 19 × 75 mm, splayed |
| 2/ | 2.08 | 4.15 | and rounded |
| | | 5.08 | 22.2.1.0.0 |

<u>width</u>
0.762

air gap b.s. say
2/0.010   0.020
          0.782

architrave b.s.
2/0.075   0.150
          0.932

<u>height</u>
1.981

air gap b.s. say
2/0.010   0.020
          2.001

architrave b.s.   0.075
              2.076

| | | | |
|---|---|---|---|
| | 0.83 | 0.83 | Painting general surfaces, ≤ 300 mm girth, internal, application on site before fixing, priming 29.1.1.1.6 (lining |
| 2/ | 2.00 | 3.99 | |
| | 0.76 | 0.76 | (door stop |
| 2/ | 1.96 | 3.92 | |
| | 0.93 | 0.93 | (architrave |
| 2/ | 2.08 | 4.15 | |
| | | 14.59 | |

Assumed not primed before delivery to site.

| | | | |
|---|---|---|---|
| | 0.83 | | Painting general surfaces, > 300 mm girth internal 29.1.2.1.0 |
| | 0.39 | 0.32 | |
| 2/ | 2.00 | | |
| | 0.39 | 1.55 | |
| | | 1.87 | |

lining girth

| | | |
|---|---|---|
| lining width | | 0.125 |
| stop depth | 2/0.019 | 0.038 |
| lining depth | 2/0.5/0.038 | 0.038 |
| architrave width | 2/0.075 | 0.150 |
| architrave depth | 2/0.019 | 0.038 |
| | | 0.389 |

| | | | |
|---|---|---|---|
| | | | ddt |
| | 0.83 | | Walls, 75 mm thick, blockwork abd |
| | 2.01 | 1.66 | |

width

| | | |
|---|---|---|
| | | 0.762 |
| lining b.s. | 2/0.032 | 0.064 |
| | | 0.826 |

height

| | |
|---|---|
| | 1.981 |
| lining | 0.032 |
| | 2.013 |

*(Continued)*

*(Continued)*

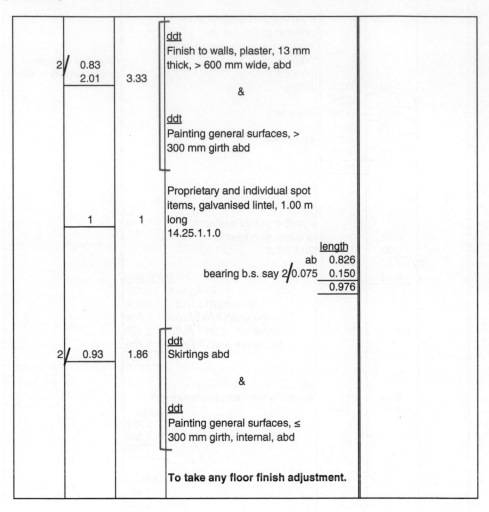

|  |  |  |  |  |
|---|---|---|---|---|
| 2/ | 0.83<br>2.01 | 3.33 | <u>ddt</u><br>Finish to walls, plaster, 13 mm<br>thick, > 600 mm wide, abd<br><br>&<br><br><u>ddt</u><br>Painting general surfaces, ><br>300 mm girth abd |  |
| 1 |  | 1 | Proprietary and individual spot<br>items, galvanised lintel, 1.00 m<br>long<br>14.25.1.1.0 |  |

|  | length |
|---|---|
| ab | 0.826 |
| bearing b.s. say 2/0.075 | 0.150 |
|  | 0.976 |

|  |  |  |  |  |
|---|---|---|---|---|
| 2/ | 0.93 | 1.86 | <u>ddt</u><br>Skirtings abd<br><br>&<br><br><u>ddt</u><br>Painting general surfaces, ≤<br>300 mm girth, internal, abd |  |

**To take any floor finish adjustment.**

# 14

# Reinforced Concrete Structures

## Generally

This type of work involves the measurement of a number of components comprising a combination of concrete, steel reinforcement, and temporary support known as *formwork*.

The measurement of a reinforced concrete structure requires a clear and logical approach to the order of the take-off so that items are not missed. The work may be subdivided into several parts of the building, and then taken off floor by floor, measuring the concrete, formwork, and reinforcement for each floor. In some instances, it may be more practical to measure all of one type of structural component, such as columns, throughout the building. It is also advisable to measure the associated formwork after the concrete so that it is not overlooked, and in any case the same dimensions are usually applicable. Again, a schedule can be of great assistance in simplifying the take-off and reducing any repetitive dimensions.

It is also normal to apply grid references to a plan to enable the different elements to be located.

## Columns

Columns should be measured between floors; in counting the number of columns, one column is taken at each grid point on every floor level. On a two-storey building, there would therefore be two columns at each point, one on the ground floor and one on the first floor. The concrete is measured in cubic metres. The formwork to simple columns is measured in square metres, taking the concreted length multiplied by the girth and stating the total number of columns for isolated columns. This allows the estimator to value the cost of forming the column kickers, which are not separately measurable, and to add an allowance per square metre to the column formwork cost.

*Willis's Elements of Quantity Surveying*, Fourteenth Edition. Roy Hills and Sandra Lee.
© 2024 John Wiley & Sons Ltd. Published 2024 by John Wiley & Sons Ltd.

## Structural floors and roofs

Structural floors and roofs are measured to the overall dimensions to the outside edge of the columns or beams. The slab concrete is measured as a cubic item, the thickness of the slab being identified less than or equal to 300 mm or as exceeding 300 mm (slabs in the same class can be grouped together).

The soffit formwork is measured in square metres to the net area excluding beam soffits and column heads. Further information is given in the description, such as the thickness of the concrete (in the stages less than or equal to 300 mm thick or over 300 mm) and the propping height less than or equal to 3 m high, over 3 m but not exceeding 4.5 m and thereafter 1.5 m stages.

Depending on the complexity of the floor plan, a typical bay size could be measured and then be multiplied by the number of bays, or the area could be measured overall and then the beam and column areas deducted. Formwork will also need to be measured to the slab edges and should be measured as a linear item in metres if less than or equal to 500 mm high, width stated and in square metres if over 500 mm in accordance with NRM2, Section 11.15.1 or 2.

## Beams

Beams should be measured between the columns. Where these are attached to concrete slabs, the volume of concrete is added to the slab concrete. This does not affect the description of the slab because it is the general thickness that is given in the description.

The formwork to the sides and soffit of the beams is measured in square metres, using the concrete length and cross-section sizes to obtain the area.

Where beam sides and floor edge coincide, the floor edge is added to the girth of beam formwork. There is no need to make any deduction from the measurements for junctions of members.

## Walls

Walls are measured in the same way as slabs, and it should be noted that they include attached columns in the same way that the beams were added to the slabs. It is common practice to cast a small part of the wall with the slab. This is called a *wall kicker*; although it makes no difference to the concrete measure, it does affect the formwork. A linear item is taken for the wall kicker and measured on the centre line of the wall; the price is to include for formwork to both sides. Formwork to single sides of walls needs to be identified and kept separate.

# Reinforcement

Bar reinforcement is measured in tonnes and grouped by nominal size as NRM2 11.33 or 11.34. It is, however, generally measured linear, and then the linear measurements are converted to a weight prior to billing by multiplying the total length by the mass per metre for the diameter of bar being measured. The calculation of the number of bars generally follows the approach used in dealing with rafters or floor joists. The distance between the first and last bars is divided by the spacing of the bars, and by adding one to the result the number of bars can be calculated. Fabric reinforcement is measured in square metres to the actual area covered, the estimator allowing in the price for the loss of material due to the laps. In practice, the required reinforcement is usually included on an engineer's schedules and therefore just needs to be 'extended'. Care should be taken to allow for concrete cover to ensure that reinforcement is below the surface of the concrete. The usual allowance is 50 mm, but this can be verified by the engineer. Mesh reinforcement is measured in square metres as NRM2 11.37.

# Approach to measurement

The measurement in Example 12 also includes for the substructure work to show how this may differ for framed buildings. The topsoil needs to be removed to the full extent of the footprint of the building; therefore, projections to cover the spread of the bases beyond the ground slab should be excavated.

## Example 12    Concrete Frame

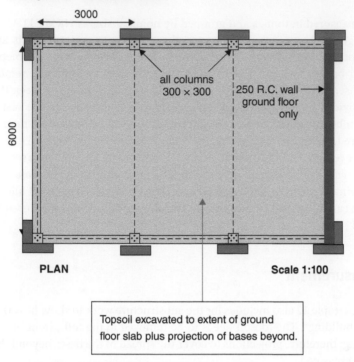

3000

6000

all columns
300 × 300

250 R.C. wall
ground floor
only

**PLAN**

**Scale 1:100**

Topsoil excavated to extent of ground
floor slab plus projection of bases beyond.

The section in Figure 36 shows how the structure is divided for measurement. Note the wall extent on the ground floor plan.

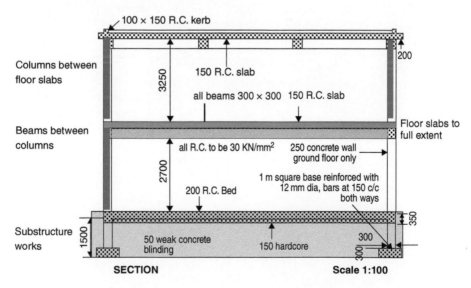

100 × 150 R.C. kerb

Columns between
floor slabs

3250

150 R.C. slab

200

all beams 300 × 300    150 R.C. slab

Beams between
columns

2700

all R.C. to be 30 KN/mm²    250 concrete wall
ground floor only

200 R.C. Bed

Floor slabs to
full extent

1 m square base reinforced with
12 mm dia, bars at 150 c/c
both ways

Substructure
works

1500

50 weak concrete
blinding

150 hardcore

350

300

300

**SECTION**

**Scale 1:100**

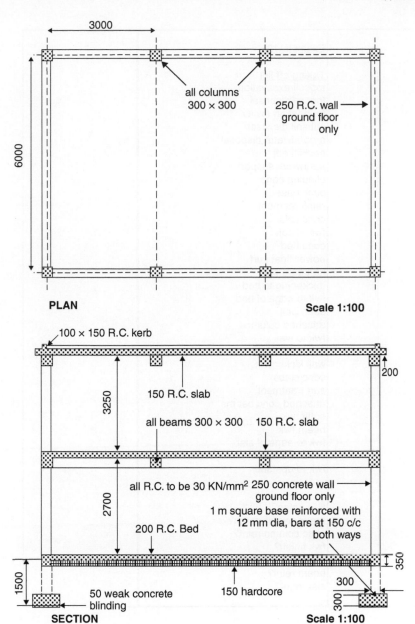

Fig. 36

Example 12

Taking-off list
topsoil excavation
topsoil disposal
foundation excav
subsoil disposal
groundwater disposal
backfill adj
earthwork support
blinding conc
conc bases
reinforcement
conc cols
fwk to cols
conc bed
power float surf
hardcore bed
thickening to bed
fwk to edge of bed
conc wall
attached columns
fwk to wall
fwk to attached cols
wall kicker fwk
conc slabs
surf treatment
attached conc beams
soffit fwk
beam fwk
fwk to edge of slab
conc upstand
fwk to upstand


RFI notes
groundwater depth?
conc column reinf?
fwk finish?
conc rf surf finish?
beam reinf?
slab reinf?

| | | | | |
|---|---|---|---|---|
| | 9.30 | | Site preparation, remove topsoil | |
| | 6.30 | 58.59 | average depth 150 mm | |
| 8/ | 1.00 | | 5.5.2.0.0 | |
| | 0.35 | 2.80 | (bases beyond slab | length |
| 4/ | 0.65 | | | 3/ 3.000 |
| | 0.35 | 0.91 | | 9.000 |
| | | 62.30 | 0.5 column b.s. | |
| | | | 2/0.5/0.300  0.300 | |
| | | | 9.300 | |

width
6.000
0.5 column b.s.
2/0.5/0.300  0.300
6.300

spread for bases
1.000
ddt col   0.300
0.5/ 0.700
0.350

base lgth
1.000
ddt spread   0.350
0.650

| | | | | |
|---|---|---|---|---|
| | 62.30 | | Retaining excavated material on site, topsoil, to temporary spoil heaps average 100 m from excavation | Using three figures shows that the quantity is m$^3$. |
| | 0.15 | | 5.10.1.1.0 | |
| | 1 | 9.35 | | |

*(Continued)*

*(Continued)*

| | | | |
|---|---|---|---|
| 8/ 1.00<br>1.00<br>1.40 | 11.20 | Excavation, starting at topsoil reduced level, foundation excavation, not exceeding 2 m deep<br>5.6.2.1.0 | |

| | |
|---|---|
| depth | 1.500 |
| conc blinding | 0.050 |
| | 1.550 |
| ddt topsoil | 0.150 |
| | 1.400 |

&

Disposal, excavated material off site
5.9.2.0.0

| | | | |
|---|---|---|---|
| | ITEM | Disposal, groundwater<br>5.9.1.0.0 | |
| 8/ 1.00<br>1.00<br>0.05 | 0.40 | Plain in situ concrete, horizontal work, ≤ 300 mm thick, in blinding, poured on or against earth or unblinded hardcore<br>11.2.1.1.1 | |
| 8/ 1.00<br>1.00<br>0.30 | 2.40 | Reinforced in situ concrete, horizontal work, ≤ 300 mm thick, in structures<br>11.2.1.2.0 | |

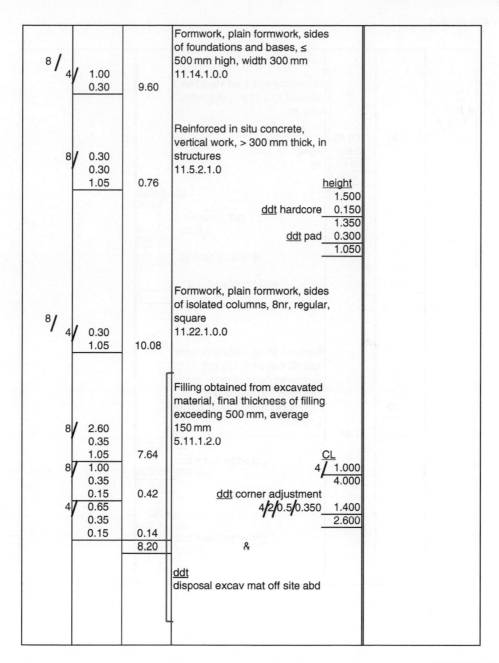

| | | | |
|---|---|---|---|
| 8/ | 4/ | 1.00 | |
| | | 0.30 | 9.60 |

Formwork, plain formwork, sides of foundations and bases, ≤ 500 mm high, width 300 mm
11.14.1.0.0

| | | | |
|---|---|---|---|
| 8/ | | 0.30 | |
| | | 0.30 | |
| | | 1.05 | 0.76 |

Reinforced in situ concrete, vertical work, > 300 mm thick, in structures
11.5.2.1.0

| | height |
|---|---|
| | 1.500 |
| ddt hardcore | 0.150 |
| | 1.350 |
| ddt pad | 0.300 |
| | 1.050 |

| | | | |
|---|---|---|---|
| 8/ | 4/ | 0.30 | |
| | | 1.05 | 10.08 |

Formwork, plain formwork, sides of isolated columns, 8nr, regular, square
11.22.1.0.0

| | | | |
|---|---|---|---|
| 8/ | | 2.60 | |
| | | 0.35 | |
| | | 1.05 | 7.64 |
| 8/ | | 1.00 | |
| | | 0.35 | |
| | | 0.15 | 0.42 |
| 4/ | | 0.65 | |
| | | 0.35 | |
| | | 0.15 | 0.14 |
| | | | 8.20 |

Filling obtained from excavated material, final thickness of filling exceeding 500 mm, average 150 mm
5.11.1.2.0

| | CL |
|---|---|
| 4/ | 1.000 |
| | 4.000 |
| ddt corner adjustment | |
| 4/2/0.5/0.350 | 1.400 |
| | 2.600 |

&

ddt
disposal excav mat off site abd

*(Continued)*

(*Continued*)

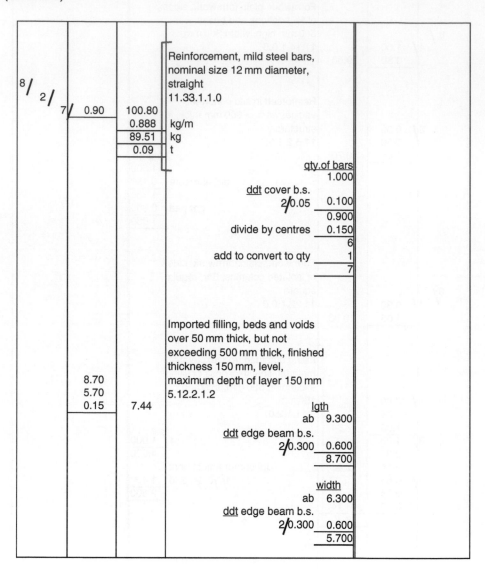

| 8/ 2/ 7/ | 0.90 | 100.80 | Reinforcement, mild steel bars, nominal size 12 mm diameter, straight 11.33.1.1.0 |
|---|---|---|---|
| | | 0.888 | kg/m |
| | | 89.51 | kg |
| | | 0.09 | t |

|  |  |
|---|---|
| qty.of bars | 1.000 |
| ddt cover b.s. 2/0.05 | 0.100 |
| | 0.900 |
| divide by centres | 0.150 |
| | 6 |
| add to convert to qty | 1 |
| | 7 |

| 8.70 5.70 0.15 | 7.44 | Imported filling, beds and voids over 50 mm thick, but not exceeding 500 mm thick, finished thickness 150 mm, level, maximum depth of layer 150 mm 5.12.2.1.2 |
|---|---|---|

|  |  |
|---|---|
| lgth ab | 9.300 |
| ddt edge beam b.s. 2/0.300 | 0.600 |
| | 8.700 |
| width ab | 6.300 |
| ddt edge beam b.s. 2/0.300 | 0.600 |
| | 5.700 |

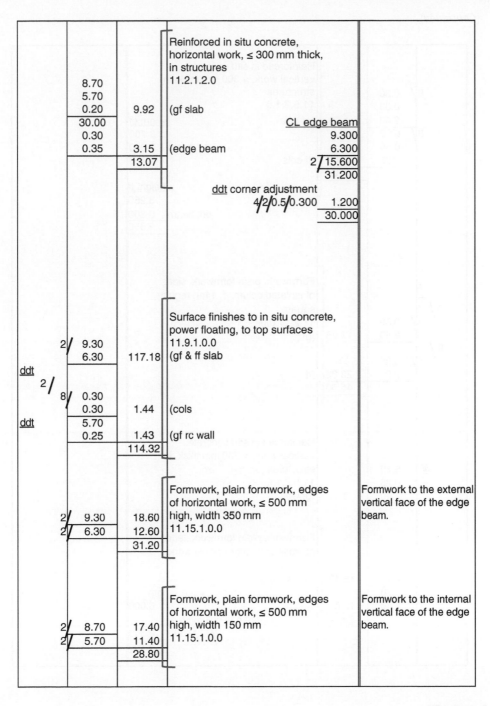

| | | | |
|---|---|---|---|
| | 8.70 | | Reinforced in situ concrete, horizontal work, ≤ 300 mm thick, in structures 11.2.1.2.0 |
| | 5.70 | | |
| | 0.20 | 9.92 | (gf slab |
| | 30.00 | | |
| | 0.30 | | |
| | 0.35 | 3.15 | (edge beam |
| | | 13.07 | |

CL edge beam
9.300
6.300
2/ 15.600
31.200

ddt corner adjustment
4/2/0.5/0.300   1.200
30.000

| | | | |
|---|---|---|---|
| 2/ | 9.30 | | Surface finishes to in situ concrete, power floating, to top surfaces 11.9.1.0.0 |
| | 6.30 | 117.18 | (gf & ff slab |
| ddt 2/ 8/ | 0.30 | | |
| | 0.30 | 1.44 | (cols |
| ddt | 5.70 | | |
| | 0.25 | 1.43 | (gf rc wall |
| | | 114.32 | |

| | | | | |
|---|---|---|---|---|
| 2/ | 9.30 | 18.60 | Formwork, plain formwork, edges of horizontal work, ≤ 500 mm high, width 350 mm 11.15.1.0.0 | Formwork to the external vertical face of the edge beam. |
| 2/ | 6.30 | 12.60 | | |
| | | 31.20 | | |

| | | | | |
|---|---|---|---|---|
| 2/ | 8.70 | 17.40 | Formwork, plain formwork, edges of horizontal work, ≤ 500 mm high, width 150 mm 11.15.1.0.0 | Formwork to the internal vertical face of the edge beam. |
| 2/ | 5.70 | 11.40 | | |
| | | 28.80 | | |

*(Continued)*

*(Continued)*

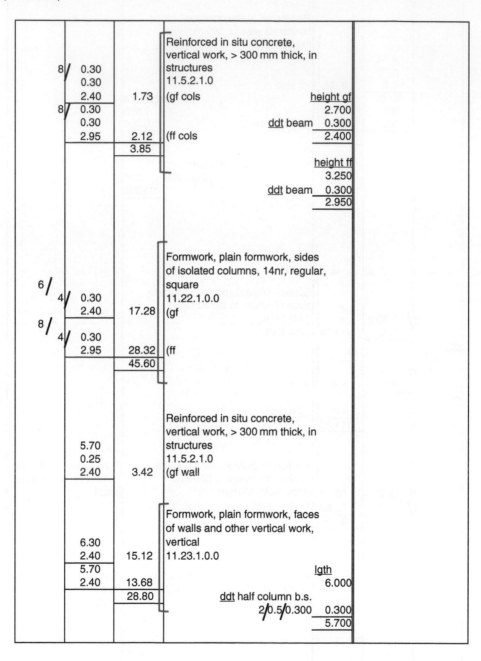

| | | | |
|---|---|---|---|
| 8/ | 0.30 | | Reinforced in situ concrete, vertical work, > 300 mm thick, in structures |
| | 0.30 | | 11.5.2.1.0 |
| | 2.40 | 1.73 | (gf cols            height gf |
| 8/ | 0.30 | |                      2.700 |
| | 0.30 | | ddt beam    0.300 |
| | 2.95 | 2.12 | (ff cols            2.400 |
| | | 3.85 | |

                                              height ff  
                                              3.250  
                       ddt beam    0.300  
                                              2.950

| | | | |
|---|---|---|---|
| 6/ | | | Formwork, plain formwork, sides of isolated columns, 14nr, regular, square |
| 4/ | 0.30 | | 11.22.1.0.0 |
| | 2.40 | 17.28 | (gf |
| 8/ | | | |
| 4/ | 0.30 | | |
| | 2.95 | 28.32 | (ff |
| | | 45.60 | |

| | | | |
|---|---|---|---|
| | 5.70 | | Reinforced in situ concrete, vertical work, > 300 mm thick, in structures |
| | 0.25 | | 11.5.2.1.0 |
| | 2.40 | 3.42 | (gf wall |

| | | | |
|---|---|---|---|
| | 6.30 | | Formwork, plain formwork, faces of walls and other vertical work, vertical |
| | 2.40 | 15.12 | 11.23.1.0.0 |
| | 5.70 | |                 lgth |
| | 2.40 | 13.68 |                  6.000 |
| | | 28.80 | ddt half column b.s. |
| | | | 2/0.5/0.300    0.300 |
| | | |                 5.700 |

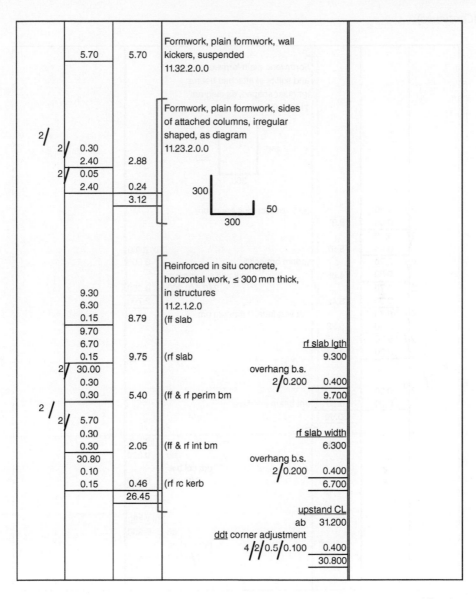

|  |  |  |  |
|---|---|---|---|
|  | 5.70 | 5.70 | Formwork, plain formwork, wall kickers, suspended 11.32.2.0.0 |

2/

|  |  |  |  |
|---|---|---|---|
| 2/ | 0.30 | | Formwork, plain formwork, sides of attached columns, irregular shaped, as diagram 11.23.2.0.0 |
| | 2.40 | 2.88 | |
| 2/ | 0.05 | | |
| | 2.40 | 0.24 | |
| | | 3.12 | |

300

300    50

Reinforced in situ concrete, horizontal work, ≤ 300 mm thick, in structures
11.2.1.2.0

|  |  |  |  |
|---|---|---|---|
| | 9.30 | | |
| | 6.30 | | |
| | 0.15 | 8.79 | (ff slab |
| | 9.70 | | |
| | 6.70 | | |
| | 0.15 | 9.75 | (rf slab |
| 2/ | 30.00 | | |
| | 0.30 | | |
| | 0.30 | 5.40 | (ff & rf perim bm |

2/

|  |  |  |  |
|---|---|---|---|
| 2/ | 5.70 | | |
| | 0.30 | | |
| | 0.30 | 2.05 | (ff & rf int bm |
| | 30.80 | | |
| | 0.10 | | |
| | 0.15 | 0.46 | (rf rc kerb |
| | | 26.45 | |

rf slab lgth
9.300
overhang b.s.
2/0.200    0.400
9.700

rf slab width
6.300
overhang b.s.
2/0.200    0.400
6.700

upstand CL
ab    31.200
ddt corner adjustment
4/2/0.5/0.100    0.400
30.800

*(Continued)*

*(Continued)*

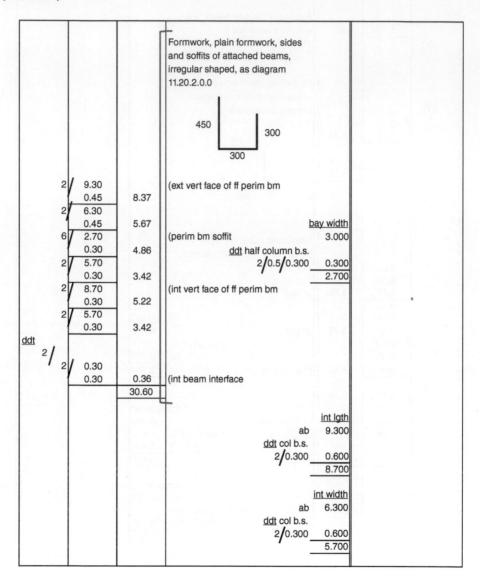

|     |       |       |                                                    |            |       |
|-----|-------|-------|----------------------------------------------------|------------|-------|
|     |       |       | Formwork, plain formwork, sides and soffits of attached beams, irregular shaped, as diagram 11.20.2.0.0 | | |
| 2/  | 9.30  |       | (ext vert face of ff perim bm                      |            |       |
|     | 0.45  | 8.37  |                                                    |            |       |
| 2/  | 6.30  |       |                                                    | bay width  |       |
|     | 0.45  | 5.67  | (perim bm soffit                                   | 3.000      |       |
| 6/  | 2.70  |       |                                                    | ddt half column b.s. | |
|     | 0.30  | 4.86  |                                                    | 2/0.5/0.300 | 0.300 |
| 2/  | 5.70  |       |                                                    |            | 2.700 |
|     | 0.30  | 3.42  | (int vert face of ff perim bm                      |            |       |
| 2/  | 8.70  |       |                                                    |            |       |
|     | 0.30  | 5.22  |                                                    |            |       |
| 2/  | 5.70  |       |                                                    |            |       |
|     | 0.30  | 3.42  |                                                    |            |       |
| ddt | 2/    |       |                                                    |            |       |
| 2/  | 0.30  |       |                                                    |            |       |
|     | 0.30  | 0.36  | (int beam interface                                |            |       |
|     |       | 30.60 |                                                    |            |       |
|     |       |       |                                                    | int lgth   |       |
|     |       |       |                                                    | ab         | 9.300 |
|     |       |       |                                                    | ddt col b.s. | |
|     |       |       |                                                    | 2/0.300    | 0.600 |
|     |       |       |                                                    |            | 8.700 |
|     |       |       |                                                    | int width  |       |
|     |       |       |                                                    | ab         | 6.300 |
|     |       |       |                                                    | ddt col b.s. | |
|     |       |       |                                                    | 2/0.300    | 0.600 |
|     |       |       |                                                    |            | 5.700 |

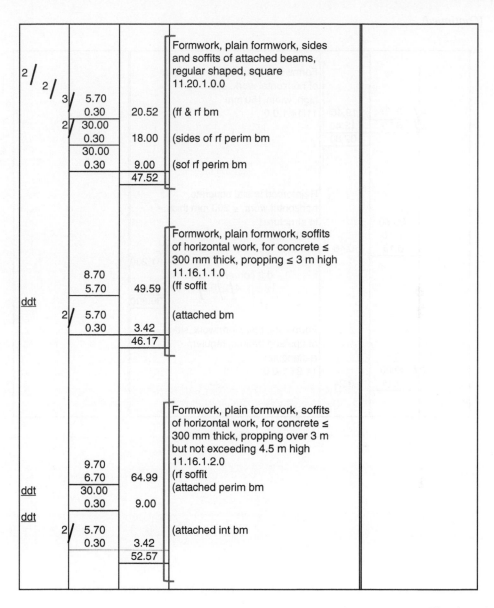

| | | | | |
|---|---|---|---|---|
| 2/ 2/ 3/ | 5.70 | | Formwork, plain formwork, sides and soffits of attached beams, regular shaped, square 11.20.1.0.0 | |
| | 0.30 | 20.52 | (ff & rf bm | |
| 2/ | 30.00 | | | |
| | 0.30 | 18.00 | (sides of rf perim bm | |
| | 30.00 | | | |
| | 0.30 | 9.00 | (sof rf perim bm | |
| | | 47.52 | | |
| ddt | 8.70 | | Formwork, plain formwork, soffits of horizontal work, for concrete ≤ 300 mm thick, propping ≤ 3 m high 11.16.1.1.0 | |
| | 5.70 | 49.59 | (ff soffit | |
| 2/ | 5.70 | | (attached bm | |
| | 0.30 | 3.42 | | |
| | | 46.17 | | |
| | 9.70 | | Formwork, plain formwork, soffits of horizontal work, for concrete ≤ 300 mm thick, propping over 3 m but not exceeding 4.5 m high 11.16.1.2.0 | |
| | 6.70 | 64.99 | (rf soffit | |
| ddt | 30.00 | | (attached perim bm | |
| | 0.30 | 9.00 | | |
| ddt | | | | |
| 2/ | 5.70 | | (attached int bm | |
| | 0.30 | 3.42 | | |
| | | 52.57 | | |

*(Continued)*

(*Continued*)

| | | | | |
|---|---|---|---|---|
| 2/ | 9.70 | 19.40 | Formwork, plain formwork, edges of horizontal work, ≤ 500 mm high, width 150 mm 11.14.1.0.0 | |
| 2/ | 6.70 | 13.40 | | |
| | | 32.80 | | |
| | 30.80 | | Reinforced in situ concrete, horizontal work, ≤ 300 mm thick, in structures 11.2.1.2.0 | |
| | 0.10 | | | |
| | 0.15 | 0.46 | <div align="right">CL<br>ab 31.200</div> | |
| | | | <div align="right"><u>ddt</u> corner adjustment<br>4/2/0.5/0.100   <u>0.400</u><br>30.800</div> | |
| 2/ | 30.80 | | Formwork, plain formwork, sides of upstand beams, regular, rectangular 11.21.1.0.0 | |
| | 0.15 | 9.24 | | |

# 15

# Structural Steelwork

The measurement of structural steel in Work Section 15 can generally be classified as:

- Framed – framing and fabrication.
- Framed – permanent erection.
- Isolated members. An isolated member is one which is not part of a frame and would cover work such as isolated beams resting on pad-stones.

The measurement of the framing is required to be split into two items, with the fabrication being kept separate from the erection, the item for erecting all of the steelwork giving the total weight involved, with the unit of measurement being tonnes.

The weight of the steel supplied to site may be greater than the calculated weight due to the manufacturing process. This is because of the rolling margin for which no allowance is made.

The supply and fabrication of the sections are generally measured by weight, and the structural function needs to be stated; therefore, you would have separate items for items such as columns, beams, bracings, trusses, and so on.

In each of the classifications, you need to have separate items for every weight class:

- Weight ≤ 25 kg per linear metre.
- Weight between 25 and 50 kg/m.
- Weight between 50 and 100 kg/m.
- And so on in 50 kg/m increments.

## Fittings

The different steel members are joined together by an assortment of plates, brackets, angles, and the like. These are called fittings, and they are measured separately by weight in a composite item of fittings for the framed members or the isolated members.

*Willis's Elements of Quantity Surveying*, Fourteenth Edition. Roy Hills and Sandra Lee.
© 2024 John Wiley & Sons Ltd. Published 2024 by John Wiley & Sons Ltd.

There are two components to a connected joint in steelwork generally:

- A fitting – for example, a cleat which is usually a short length of steel section, either an angle or channel which acts as the connecting agent and is measured as described in this chapter.
- A fixing – this could be rivets, bolts, or distance pieces.

The fixing devices are as follows:

- Black bolts – forged from round stock with only the threads machined, and used in clearance holes about 1 mm larger than the diameter of the bolt. These are deemed included in the weight of structural members.
- Turned bolts – machined on shank and under head to fit holes with a very small clearance. These are enumerated as 15.11.1.
- Friction grip bolts – high-tensile steel that is tightened to a predetermined tension so that the load is carried by friction; nowadays, this is often used instead of site riveting. Measured as turned bolts using the description special bolts.

Holes for bolts and so on are deemed included in the contractors' prices.

Fixing bolts for isolated members are measured where they are required for fixing to other elements and would be enumerated.

Offsite and onsite surface preparation and treatments are measured in square metres.

The measurement of steel framed structures is no more difficult than that of any other type of construction but, similar to other areas, the following of a logical approach is essential as it is easy to make errors in counting or to omit whole sections of the work.

One approach would be to follow an order of dealing with all work on a particular floor before commencing the next floor. By following this order, there is less chance of something being missed. Within each floor, you might follow the following order.

*Main order*

- Columns
- Main beams
- Secondary beam

  - Filler beams
  - Compound trusses
  - Purlins
  - Rails
  - Braces
  - Struts

| Main member measurement order | Fittings item |
|---|---|
| • Columns (full length taken) | Base plates |
| | Splices |
| | Caps |
| | Cleats |
| • Beams | Shelf angles |
| | Cleats |
| | Gusset plates |
| | Angles |
| • Truss main members | |
| • Struts | |
| • Ties | |

Looking at this order, it will be seen that the columns are given priority of measure, followed by the beams, and so on. The fittings order extends the measure of the individual members of the first order to deal with the constituent parts. Splay cuts to members or plates are not measurable.

The measurement of concrete casing to beams and columns would be measured under 11.2.2.2 and 11.5.2.1, respectively, and the associated formwork as 11.19 to 11.23 as appropriate. Sprayed in situ concrete would be measured under 11.7.

Refer to Example 13 for the measure of a simple steel frame.

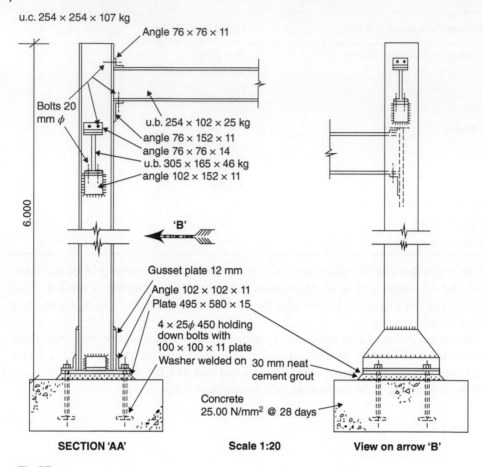

u.c. 254 × 254 × 107 kg

Angle 76 × 76 × 11

Bolts 20 mm $\phi$

u.b. 254 × 102 × 25 kg
angle 76 × 152 × 11
angle 76 × 76 × 14
u.b. 305 × 165 × 46 kg
angle 102 × 152 × 11

6.000

'B'

Gusset plate 12 mm

Angle 102 × 102 × 11

Plate 495 × 580 × 15

4 × 25$\phi$ 450 holding
down bolts with
100 × 100 × 11 plate
Washer welded on

30 mm neat
cement grout

Concrete
25.00 N/mm² @ 28 days

**SECTION 'AA'**          **Scale 1:20**          **View on arrow 'B'**

**Fig. 37**

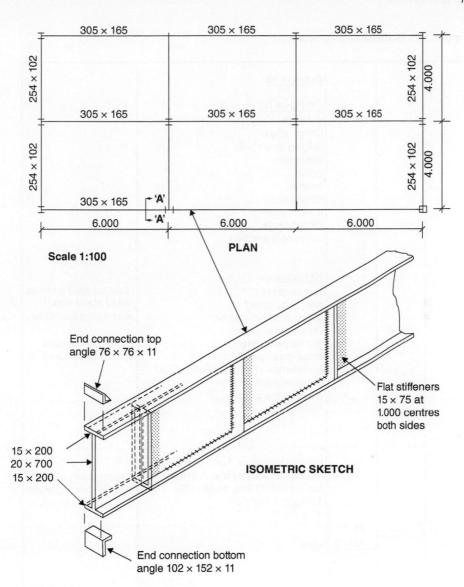

305 × 165    305 × 165    305 × 165

254 × 102    254 × 102    4.000

305 × 165    305 × 165    305 × 165

254 × 102    254 × 102    4.000

305 × 165    'A'

6.000    6.000    6.000

**Scale 1:100**    **PLAN**

End connection top
angle 76 × 76 × 11

Flat stiffeners
15 × 75 at
1.000 centres
both sides

15 × 200
20 × 700
15 × 200

**ISOMETRIC SKETCH**

End connection bottom
angle 102 × 152 × 11

**Fig. 37 (Cont'd)**

**Example 13   Structural Steelwork**

| | | | | |
|---|---|---|---|---|
| | | | Example 13 | |
| | | | Taking-off list<br>columns<br>fittings angle<br>holding down bolts<br>baseplate<br>grout<br>beams<br>fittings angle<br>builtup girder<br>fittings angle<br>permanent erection | |
| | | | RFI notes<br>surf treatment?<br>baseplate dims?<br>girder midpoint column?<br>$m^3$ weight of steel?<br>gusset plate dims?<br>angle mass per m?<br>col orientation at girder connection?<br>girder midpoint beam?<br>surf treatment?<br>testing? | Note: labelled gridlines would assist with referencing quantities.<br><br>Note: non-standard steel sections used. |
| | | | Framed members, framing and fabrication, lengths over 1 m but not exceeding 9 m, weight 100–150kg/m, columns<br>15.1.2.4.1 | Assumed no column required to support the middle of the girder. |
| 11/ | 5.96 | 65.51<br>107.00<br>7009.04<br>7.01 | kg/m<br>kg<br>t | |

col 254×254

height   6.000
ddt grout   0.030
   5.970
ddt baseplate   0.015
   5.955

| | | | | |
|---|---|---|---|---|
| 11/ | 0.50 | | Allowance for fitting, calculated weight, to framed members 15.5.1.1.0 | |
| | 0.58 | | | |
| 11/ | 0.02 | 0.05 | (baseplate | Rounded thickness value appears in spreadsheet. |
| 2/ | 0.58 | | (gusset plate | |
| | 0.30 | | | |
| | 0.01 | 0.05 | | |
| | | 0.09 | $m^3$ grout 0.030 | |
| | | 7854.00 | $kg/m^3$ plate 0.015 | |
| | | 736.60 | kg | |
| | | 0.74 | t | |

| | | | | |
|---|---|---|---|---|
| | | | Allowance for fitting, calculated weight, to framed members 15.5.1.1.0 | |
| 11/ | | | | |
| 2/ | 0.58 | 12.76 | (baseplate angle 102×102×11 flange | |
| 11/ | | | | |
| 2/ | 0.15 | 3.30 | (baseplate angle 102×102×11 web | |
| | | 16.06 | | |
| | | 16.00 | kg/m | |
| | | 256.96 | kg | |
| | | 0.26 | t | |

| | | | | |
|---|---|---|---|---|
| | | | Allowance for fitting, calculated weight, to framed members 15.5.1.1.0 | |
| 7/ | | | | |
| 2/ | 0.17 | 2.38 | (102×152×11 angle for 305×165 bm | |
| | | 26.60 | kg/m | |
| | | 63.31 | kg | |
| | | 0.06 | t | |

*(Continued)*

*(Continued)*

| | | | | |
|---|---|---|---|---|
| 8 / | | | | Allowance for fittings, calculated weight, to framed members 15.5.1.1.0 |
| | 2 / | 0.10 | 1.60 | (76 × 152 × 11 angle for 254 × 102mm bm |
| | | | 20.20 | kg/m |
| | | | 32.32 | kg |
| | | | 0.03 | t |
| 7 / | | | | Allowance for fittings, calculated weight, to framed members 15.5.1.1.0 |
| | 2 / | 0.17 | 2.38 | (76 × 76 × 14 angle for 305 × 165mm bm |
| | | | 15.90 | kg/m |
| | | | 37.84 | kg |
| | | | 0.04 | t |
| 7 / | | | | Allowance for fittings, calculated weight, to framed members 15.5.1.1.0 |
| | 2 / | 0.10 | 1.40 | (76 × 76 × 11 angle for 254 × 102mm bm |
| | | | 13.40 | kg/m |
| | | | 18.76 | kg |
| | | | 0.02 | t |
| | | 11 | 11 | Holding down bolts or assemblies, 4nr 25mm diameter bolts 450mm long with 100 × 100 × 11mm plate washer 15.10.1.1.0 |
| | | | | & |
| | | | | In situ concrete sundries, grouting, 495 × 580mm, stanchion bases 11.42.1.1.0 |

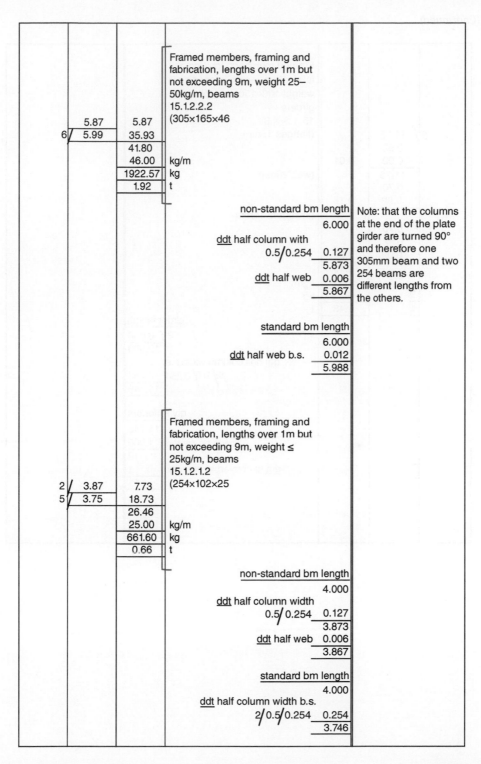

|  |  |  |  |
|---|---|---|---|
| | | | Framed members, framing and fabrication, lengths over 1m but not exceeding 9m, weight 25–50kg/m, beams 15.1.2.2.2 (305×165×46 |
| 6/ | 5.87 5.99 | 5.87 35.93 | |
| | | 41.80 | |
| | | 46.00 | kg/m |
| | | 1922.57 | kg |
| | | 1.92 | t |

non-standard bm length
6.000
ddt half column with
0.5/0.254   0.127
5.873
ddt half web   0.006
5.867

standard bm length
6.000
ddt half web b.s.   0.012
5.988

Note: that the columns at the end of the plate girder are turned 90° and therefore one 305mm beam and two 254 beams are different lengths from the others.

|  |  |  |  |
|---|---|---|---|
| | | | Framed members, framing and fabrication, lengths over 1m but not exceeding 9m, weight ≤ 25kg/m, beams 15.1.2.1.2 (254×102×25 |
| 2/ 5/ | 3.87 3.75 | 7.73 18.73 | |
| | | 26.46 | |
| | | 25.00 | kg/m |
| | | 661.60 | kg |
| | | 0.66 | t |

non-standard bm length
4.000
ddt half column width
0.5/0.254   0.127
3.873
ddt half web   0.006
3.867

standard bm length
4.000
ddt half column width b.s.
2/0.5/0.254   0.254
3.746

*(Continued)*

*(Continued)*

| | | | | |
|---|---|---|---|---|
| | | | Framed members, framing and fabrication, length exceeding 9m, weight 100–150kg/m, plate girders<br>15.1.3.X.8<br>(flanges 15mm | |
| 2/ | 11.75<br>0.02<br>0.02 | 0.01 | | |
| | 11.75<br>0.70<br>0.02 | 0.16 | (web 20mm | |
| 2/<br>11/ | 0.70<br>0.08<br>0.02 | 0.02 | (stiffeners 15mm | |
| | | 0.19<br>7854.00<br>1482.97<br>1.48 | kg/m$^3$<br>kg<br>t | |

girder length
2/ 6.000
12.000

ddt half column width b.s.
2/ 0.5/ 0.254   0.254
11.746

qty stiffeners
11.746
divide by centers   1.000
12
ddt to convert to quantity   1
11

weight/m
126.25

| | | | |
|---|---|---|---|
| | | | Allowance for fittings, calculated weight, to framed members 15.5.1.1.0 |
| 2/ | 0.20 | 0.40 | (76×76×11 angle for girder bm |
| | | 13.40 | kg/m |
| | | 5.36 | kg |
| | | 0.01 | t |
| | | | |
| | | | Allowance for fittings, calculated weight, to framed members 15.5.1.1.0 |
| 2/ | 0.20 | 0.40 | (102×152×11 angle for girder bm |
| | | 26.60 | kg/m |
| | | 10.64 | kg |
| | | 0.01 | t |
| | | | |
| | | | Framed members, permanent erection on site 15.2.0.0.0 |
| | 7.01 | 7.01 | (254×254×107 cols |
| | 1.92 | 1.92 | (305×165×46 bm |
| | 0.56 | 0.56 | (254×102×25 bm |
| | 1.48 | 1.48 | (girder |
| | | 10.98 | |

# 16

# Plumbing

## Subdivision

When measuring plumbing, it is particularly important to follow a logical sequence of taking-off in order to be sure that no part is missed. Frequently, particularly on a small domestic installation, the only information shown on the drawings is the location of sanitary appliances. If this is the case, then the measurement of the appliances is fairly straightforward and forms a logical start.

After the sanitary appliances have been measured and possibly coloured in on the drawings, it is then easier to decide on a pipework layout for both wastes and supplies. Sizes of waste pipes are dictated by the size of the waste fitting from the appliance (e.g. wash hand basins 32 mm and baths and sinks 38 mm). Supply pipework sizes for a small installation should not be too difficult to assess. The rising main is usually in 15 mm pipework, and the down feeds from the cistern in 28 or 22 mm, reducing to 15 mm for the individual feeds except for baths, which require 22 mm. Adequate isolating and drain-off valves should be included in the system to enable sections to be isolated and drained. Any assumptions should be clarified with the appropriate designer.

Gate valves and ball valves do not restrict the flow of water when fully open and should be used on the low-pressure distribution part of the system. Some water supply companies still require an indirect system, with a drinking feed to the kitchen sink taken off the rising main and the other cold feeds coming from a storage tank. The capacity of the cold water tank is given as either nominal (i.e. filled to the top edge) or actual (i.e. filled to the working water line). The requirements of water supply companies vary considerably in required capacities, but this should be at least 112 L actual for storage only, rising to 225 L for storage and feeding a hot water system. The cold water tank may have to be raised to provide adequate pressure and flow, particularly for showers, and the roof construction may have to be strengthened to support the additional weight. The inlet to the tank is controlled by a ballvalve, and the outlet should be opposite the inlet to avoid stagnation of water. An overflow pipe with twice the capacity of the inlet should be provided. An alternative installation, where permitted, is for all appliances to be fed directly from the incoming main supply, without the need for a cold water tank.

Before attempting to measure a plumbing installation, trade catalogues depicting fittings available for the specified pipework should be obtained for reference. A selection can then be made of suitable fittings for connections to various appliances. A diagrammatic layout (i.e. schematic) of the plumbing when provided is often not to scale and drawn in two dimensions. When measuring from such a diagram, one has to visualise the layout in three dimensions and to relate pipe runs to the structure. This will enable realistic lengths of pipes to be measured and the correct number of bends to be taken. Sometimes, when measuring copper pipes, it is difficult to decide whether to take *made* bends (i.e. the pipe bent to form the bend) or fittings. Generally, made bends should only be taken for minor changes in the direction of pipes or on short lengths. It should be remembered that long lengths of pipes with made bends may be impossible to install. For the measurement of installations without detailed information, the following is suggested as a suitable order:

| | |
|---|---|
| Sanitary appliances | (a) Sanitary appliances, including taps, traps, brackets, waste, and overflow fittings |
| Foul drainage above ground | (b) Wastes, overflow pipes, soil, and ventilating pipes, including ducts |
| Cold water installation | (c) Connection to supply company's main, supply to boundary of site and meter/stopvalve pit, reinstatement of highway |
| | (d) Supply in trench from boundary to building, stopvalve, and rising main to the cold water tank |
| | (e) Branches from rising main, including exterior taps and non-return valves |
| | (f) Cold water tank and lid, including bearers, overflow, and insulation |
| | (g) Cold down services |
| Hot water installation | (h) Feed from cold water tank |
| | (i) Boiler, flue, controls, and work in connection |
| | (j) Cylinder and primary flow and return pipes |
| | (k) Secondary circulation, expansion pipe, and branch services |
| Generally | (l) Casings |
| | (m) Chlorination, testing, and so on |

*Note*: Insulation of pipes and builder's work in connection (e.g. chases, holes, and painting) should be taken after each subdivision. Testing and commissioning should also be considered.

An alternative approach to measurement is to follow the flow of water from the water main, through the building to the sanitary appliances and discharging into the drains. This is a more logical approach and would probably be adopted if the layout of the whole system is shown on the drawings. The main divisions shown here could still be used but in a different order.

As far as presentation in the bill is concerned, plumbing work has to be classified under headings indicating the nature of the work. For a simple domestic type installation, these would be as follows:

a) Sanitary appliances.
b) Foul drainage above ground.
c) Cold water.
d) Hot water.
e) Sundry builder's work in connection with services.
f) Testing and commissioning the drainage or water system.

*Note*: For small-scale installations, (c) and (d) may be combined.

The builder's work in connection section may be billed either under a heading at the end of each appropriate work section or after the installation measurement.

When taking-off, one obviously has to keep in mind these bill divisions but by following the suggested order of measurement given earlier, the necessary sections will be automatically produced.

## Sanitary appliances

If sanitary appliances are specified fully, then they are enumerated as described in Work Section 32 and the description should include the type, colour, size, capacity, and method of fixing, including details of supports, mountings, and bedding and pointing. Frequently, a catalogue or building supply reference is used for part of the description, but care must be taken to define any alternatives available. If full details are not available, then a prime cost sum may be included for the supply of the appliances, an item included for contractor's profit and fixing measured as enumerated items. Descriptions should make it clear whether items such as taps, traps, overflow assemblies, and bath panels are included with the appliance. Small items such as towel rails, mirrors, and soap dishes must not be overlooked. Any builder's work such as tile splashbacks, bearers, backboards, painting, and similar items necessary for the installation should be measured at this stage.

## Foul drainage above ground

Included in Work Section 33 is the measurement of waste pipes, overflow pipes, anti-siphonic pipes, and soil and ventilating pipes. Some appliances, such as water closets (w.c.s), have traps built in, and some, such as wash basins, have integral overflows. Traps and other pipework ancillaries are enumerated with a dimensioned description and the method of jointing stated, although cutting pipes and jointing materials are deemed to be included. Nowadays, most waste and soil pipes are specified to be in plastic, and their description should state whether they have ring seal or solvent-welded joints and the type and spacing of pipe supports. Pipes are measured linear over fittings, and joints in the running length (i.e. jointing straight lengths of pipe together) are deemed included. The nominal size of pipes has to be given; copper and plastic are usually described by their external diameter, and cast or spun iron and mild steel by their nominal bore. Straight and curved pipes have to be classified separately, and in the case of the latter, the radius should be stated. Fixing the pipes to special backgrounds is given as follows:

- To timber, including manufactured building boards.
- To masonry, which is deemed to include concrete, brick, block, and stone.
- To metal.
- To metal-faced material.
- To vulnerable materials, which are deemed to include glass, marble, mosaic, tiled finishes, or similar.

The location of the pipework needs to be included in the description, whether it is in roofs; high or low level in plant rooms, risers, or service ducts; or high or low level on floors or in trenches.

Pipework ancillaries are enumerated and described. Fittings are enumerated and classified as ≤ 65 mm or > 65 mm and also distinguished by shape, i.e. one end, two ends, three ends etc.

General builder's work in connection with an installation is included as an item, along with marking the position of holes, mortises, and chases. Pipe and duct sleeves are enumerated with the size and type stated. Painting pipes and radiators, and other painting services, are measured linear to pipes not exceeding 300 mm girth and superficial to those exceeding 300 mm girth. The measurement of overflow pipes to flushing cisterns must not be overlooked; these are measured in the same way as waste pipes.

## Cold water

The measurement of the cold water installation under Work Section 38 will start invariably with the connection to the supply company's main; this would in all probability be included as a provisional sum. It is necessary to check with the company the extent of the work that they will carry out. Often included with the connection is the pipework to the meter/stopvalve pit at the boundary and making good the highway. The meter/stopvalve is required to be located on the pavement or just inside the boundary. Stopvalves, gatevalves, and ballvalves are defined as pipework ancillaries, enumerated and described, and their method of jointing stated. If sufficient detail exists, fittings should be identified separately. If there is insufficient detail, then the fittings may be measured inclusive of fittings. Storage tanks are enumerated as primary equipment and are described, including the size and capacity. Overflows to tanks should be taken at this stage.

Insulation to pipelines is measured linear and described, including the thickness of the insulation and the nominal size of the pipe. Working insulation around ancillaries is enumerated as extra over the insulation. Insulation to equipment is either measured superficial (on the surface of the insulation) or enumerated giving the overall size. In the former case, working around ancillaries is enumerated and in the latter case can be included in the item description. Excavating trenches for services are measured linear, stating the average depth in 500 mm stages. Earthwork support, consolidation, backfilling, and disposal are deemed to be included in the trench item. Meter/stopvalve chambers and boxes are each enumerated and described. Underground ducts are measured as linear items, stating the type, nominal size, and method of jointing, and whether they are straight or curved. The remainder of the builder's work is measured as described in this book.

Refer to Example 14 for the measure of a simple domestic installation of sanitary appliances, the associated water disposal, and the cold water supplies.

## Hot water

Traditional domestic hot water systems, apart from the pipework, have three main components – the boiler, the cylinder for storage, and the cold feed storage. Suitable pipe sizes would be 28 mm for the primary flow and return between the cylinder and boiler and for the cold feed to the cylinder. The hot water distribution from the cylinder would be 28 mm, reducing to 22 mm for the vent, 15 mm for sink or basin supplies, and 22 mm for the bath. These sizes should be regarded as minima; sizes would be increased for a larger number of draw-off points. When an indirect heating circuit is included in the system, then either a self-venting cylinder or a separate expansion and feed tank have to be provided. Whilst heating installations are considered to be beyond the scope of this book, it is worth mentioning that a separate bill heading of low-temperature hot water heating (small-scale) would have to be introduced. Boilers and cylinders are enumerated and described under the rules for equipment; the description should include, as appropriate, the type, size, pattern, rated duty, capacity, loading, and method of jointing. The remainder of the work is measured as described here.

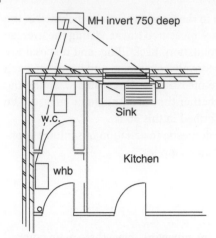

**GROUND FLOOR PLAN**

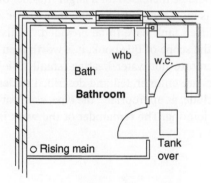

**FIRST FLOOR PLAN**

**Scale 1:100**

**Fig. 38**

**Example 14    Internal Plumbing**

| | | | | Example14 | |
|---|---|---|---|---|---|
| | | | | Taking-off list | |
| | | | | FFE | FFE = Furniture, fittings, and equipment |
| | | | | sink and drainer | |
| | | | | wc | |
| | | | | whb | |
| | | | | bath | |
| | | | | pc sum for s.o. | |
| | | | | | |
| | | | | FW | |
| | | | | 110 mm pipes | |
| | | | | bends | |
| | | | | 32 mm trap | |
| | | | | 40 mm trap | |
| | | | | 32 mm pipe | |
| | | | | 40 mm pipe fittings | |
| | | | | overflow pipe & fittings | |
| | | | | marking holes | |
| | | | | bwic | bwic = builder's work in connection |
| | | | | testing & commissioning | |
| | | | | | |
| | | | | CW | |
| | | | | primary equip – tank | |
| | | | | pipework | |
| | | | | fittings | |
| | | | | ancillaries - valves | |
| | | | | overflow fittings | |
| | | | | marking holes | |
| | | | | bwic | |
| | | | | testing & commissioning | |
| | | | | | |
| | | | | sundries | |
| | | | | insulation to pipework | |
| | | | | ditto to tank | |
| | | | | paint pipework | |
| | | | | tank bearers | |
| | | | | tank platform | |
| | | | | | |
| | | | | RFI notes | |
| | | | | spec for sanitary appliances? | |
| | | | | svp duct spec? | |
| | | | | flr to clg heights? | |
| | | | | svp terminal spec? | |
| | | | | boss branch/waste water detail? | |
| | | | | pipework lengths? | |
| | | | | svp roof sleeve detail? | |
| | | | | water tank support detail? | |
| | | | | FW & CW system schematics? | |

*(Continued)*

(*Continued*)

| | | | | |
|---|---|---|---|---|
| | Item | | **Sanitary Appliances**<br><br>Include the Prime Cost sum of £xxx for the supply only of sanitary appliances. | Assumes that the specification has not yet been decided. |
| | Item | | Allow for over heads and profit xx% on same. | |
| | | | Fix only the following furniture, fittings, and equipment. | |
| | 2 | 2 | Fixtures, fittings, or equipment with services, white glazed vitreous china wash hand basin including taps, waste fitting, plug and chain, pedestal, fixed as specification 32.2.0.1.1 | |
| | 2 | 2 | Fixtures, fittings, or equipment with services, white glazed vitreous china wc suite including cistern and ball valve, plastic flush pipe, seat and collar, fixed as specification 32.2.0.1.1 | |
| | 1 | 1 | Fixtures, fittings, or equipment with services, white reinforced bath size 1700 × 700 mm including taps, overflow and waste fittings, plug and chain, moulded side panels, fixed as specification 32.2.0.1.1 | |

| | | | | |
|---|---|---|---|---|
| 1 | 1 | Fixtures, fittings, or equipment with services, stainless steel single bowl drainer sink with taps, overflow and waste fittings, plug and chain, fixed as specification 32.2.0.1.1 | | |
| 1 | 1 | Fixtures, fittings, or equipment with services, toilet roll holder, fixed as specification 32.2.0.1.1 | | |
| 2.40 0.60 | 1.44 | Backing and other first fix timbers, nominal dimension 38 × 38 mm, framed battens, 60 mm centres 16.3.2.1.1 | | Bath panel framing. |

| | lgth |
|---|---|
| bath lgth | 1.700 |
| bath width | 0.700 |
| | 2.400 |

**Foul drainage installations**

| | | | |
|---|---|---|---|
| 6.50 | 6.50 | Pipework, uPVC, nominal diameter 110mm, straight, fixed as specification 33.1.1.1.1 | |
| 0.80 | 0.80 | (gf wc branch | |
| 0.30 | 0.30 | (ff wc branch | |
| | 7.60 | | |

| | lgth |
|---|---|
| gf | 2.600 |
| floor | 0.250 |
| ff | 2.600 |
| roof space | 0.600 |
| above roof | 0.450 |
| | 6.500 |

*(Continued)*

*(Continued)*

| | | | |
|---|---|---|---|
| 1 | 1 | Pipework ancillaries, 110 mm terminal cover 33.2.1.0.0 | |
| 2 | 2 | Items extra over the pipe in which they occur, fittings, nominal pipe size > 65 mm, two ends, wc connector 33.3.2.2.0 | |
| 2 | 2 | Items extra over the pipe in which they occur, fittings, nominal pipe size > 65 mm, three ends, boss branch 33.3.2.2.0 | |
| 2 | 2 | Pipework ancillaries, 32 mm P trap 33.2.1.0.0 (whb trap | |
| 1 | 1 | Pipework ancillaries, 40 mm P trap 33.2.1.0.0 (sink trap | |
| 1 | 1 | Pipework ancillaries, 40 mm bath trap with overflow 33.2.1.0.0 | |

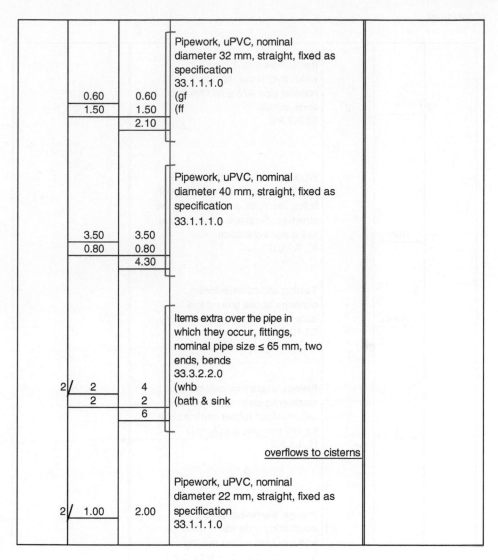

| | | | |
|---|---|---|---|
| | 0.60 | 0.60 | Pipework, uPVC, nominal diameter 32 mm, straight, fixed as specification 33.1.1.1.0 (gf |
| | 1.50 | 1.50 | (ff |
| | | 2.10 | |
| | | | |
| | 3.50 | 3.50 | Pipework, uPVC, nominal diameter 40 mm, straight, fixed as specification 33.1.1.1.0 |
| | 0.80 | 0.80 | |
| | | 4.30 | |
| 2/ | 2 | 4 | Items extra over the pipe in which they occur, fittings, nominal pipe size ≤ 65 mm, two ends, bends 33.3.2.2.0 (whb |
| | 2 | 2 | (bath & sink |
| | | 6 | |

<u>overflows to cisterns</u>

| | | | |
|---|---|---|---|
| 2/ | 1.00 | 2.00 | Pipework, uPVC, nominal diameter 22 mm, straight, fixed as specification 33.1.1.1.0 |

*(Continued)*

(Continued)

| | | | | |
|---|---|---|---|---|
| 2/ 1 | 2 | Items extra over the pipe in which they occur, fittings, nominal pipe size ≤ 65 mm, two ends, bends 33.3.2.2.0 | | |
| Item | | Work for services installations in new buildings, marking position of holes, mortices, and chases in the structure, drainage above ground foul water installation 41.2.1.0.0 | | |
| Item | | Testing and commissioning, drainage above ground foul water installation 33.10.1.0.0 | | |
| 1 | 1 | Fittings, aluminium patent weathering slate 457 × 400 mm with moulded rubber sealing cone for 110 mm svp, supply only 18.4.1.8.2<br><br>&<br><br>Fittings, aluminium patent weathering slate 457 × 400 mm with moulded rubber sealing cone for 110 mm svp, fix only by roofer | | |

| | | | cold water supply system | |
|---|---|---|---|---|
| | | | Pipework, copper, nominal diameter 15 mm, fixed to masonry, riser | |
| 5.45 | 5.45 | | 38.3.1.1.0 | |
| 7.80 | 7.80 | | riser | |
| 9.10 | 9.10 | | gf 2.600 | |
| | 22.35 | | floor 0.250 | |
| | | | ff 2.600 | |
| | | | 5.450 | |
| | | | | |
| | | | gf | |
| | | | 3.200 | Pipework, scaled from |
| | | | 2.500 | drawings. |
| | | | to sink 0.700 | |
| | | | to wc 0.700 | |
| | | | to whb 0.700 | |
| | | | 7.800 | |
| | | | | |
| | | | ff | |
| | | | to wc 3.200 | |
| | | | 3.800 | |
| | | | 0.700 | |
| | | | to whb 0.700 | |
| | | | to bath 0.700 | |
| | | | 9.100 | |
| | | | | |
| | | | Pipework, copper, nominal diameter 15 mm, fixed to timber | |
| 4.60 | 4.60 | | high level in roof space | |
| | | | 38.3.1.1.0 | |
| | | | tank in rf space | |
| | | | 4.000 | |
| | | | clg 0.100 | |
| | | | vent to cistern 0.500 | |
| | | | 4.600 | |

(*Continued*)

*(Continued)*

| | | | | |
|---|---|---|---|---|
| 1 | 1 | Pipe ancillaries, stop valve, and drain tap<br>38.5.1.0.0 | |
| 1 | 1 | Primary equipment, cold water supply system, high level in roof space, water storage tank 225L including ball valve, connection coupling and insulating jacket<br>38.1.1.1.1 | |
| 3.50 | 3.50 | Pipework, uPVC, nominal diameter 22 mm, fixed to timber, in roof space<br>38.3.1.1.0<br>(overflow | |
| Item | | Testing<br>33.17.0.0.0 | |
| Item | | Commissioning<br>33.17.0.0.0 | |
| Item | | Work for services installations in new buildings, marking position of holes, mortices, and chases in the structure, cold water installation<br>41.2.1.0.0 | |

| | | | |
|---|---|---|---|
| | 5.45 | 5.45 | Painting general services, ≤ 300 mm girth, internal<br>29.1.1.1.0<br>(exposed pipework |
| | 7.80 | 7.80 | |
| | 9.10 | 9.10 | |
| | | 22.35 | |
| 3/ | 1.50 | 4.50 | Primary or structural timbers, nominal size 50 × 100 mm, tank bearer, fixed to timber<br>16.1.1.5.6 |
| | 1.50 | | Boarding, over 600 mm wide, finished thickness 19 mm, horizontal<br>16.4.2.1.0<br>(water tank base |
| | 1.25 | 1.88 | |
| | 4.60 | 4.60 | Insulation and fire protection, 20 mm thick to 15 mm copper pipe, to pipework<br>38.9.2.0.1 |

# 17

# Electrical Services

Generally, the measurement of electrical services under Work Section 39 may be divided into the following categories:

- Primary equipment which is enumerated.
- Terminal equipment and fittings which are also enumerated.
- Cables measured in metres.

Different electrical systems, i.e. lighting and power, should be separately measured and preceded by an appropriate title to guide the estimator. The information that should be provided includes floor plans, sections, and elevations as the system layout will be indicated on schematics which do not show the exact position of equipment in the structure.

Larger commercial installations will usually require the measured work such as cables, to be identified by the use of labels, etc., and this work is separately measurable as an item. Fire stopping is enumerated and will be required where services pass through fire compartments. Testing, commissioning, and the provision of operation and maintenance manuals are all measurable as items. It may also be necessary to provide record drawings of where the services are actually installed within the structure and this too is measurable as an item.

Builder's work in connection with electrical installations is measurable under Work Section 41 although the marking of holes, chases, and mortices is deemed to be included in Work Section 39. Where external services are being measured to trenches, then these too are measured under Work Section 41 using rules comparable to those used to measure underground drainage.

When measuring services, it is useful to prepare a schedule on a spreadsheet which shows where equipment is located as this may then be used to quickly and accurately determine totals.

*Willis's Elements of Quantity Surveying*, Fourteenth Edition. Roy Hills and Sandra Lee.
© 2024 John Wiley & Sons Ltd. Published 2024 by John Wiley & Sons Ltd.

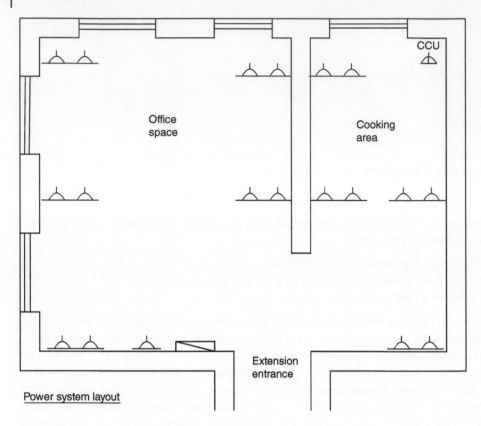

Office space

Cooking area

CCU

Extension entrance

Power system layout

Key

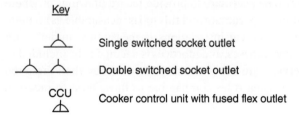

Single switched socket outlet

Double switched socket outlet

CCU
Cooker control unit with fused flex outlet

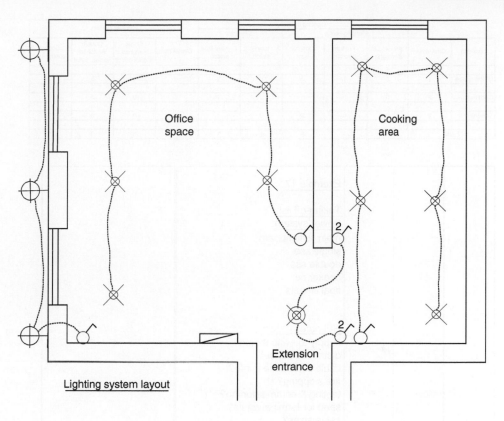

Office
space

Cooking
area

Extension
entrance

Lighting system layout

## Key

⊗ Ceiling light – recessed LED downlighter

⊚ Ceiling light – feature pendant

▱ Main switchboard

⊕ External light

2⌒ Wall – mounted 2 – way light switch

⌒ Wall – mounted 1 – way light switch

| Location | Circuit no. | Lighting system | | | | | Power system | | |
| | | Lighting points | | | Switches | | Circuit no. | Single switched socket outlet | Double switched socket outlet | Cooker control unit |
| | | Recessed LED downlighter | Feature pendant | External light | 2 – way light switch | 1 – way light switch | | | | |
|---|---|---|---|---|---|---|---|---|---|---|
| Office space | 1 | 5 | 1 | 0 | 0 | 1 | 4 | 1 | 5 | 0 |
| Cooking area | 2 | 5 | 0 | 0 | 2 | 1 | 4 | 0 | 4 | 0 |
| External | 3 | 0 | 0 | 3 | 0 | 1 | 5 | 0 | 0 | 1 |
| Totals | | 10 | 1 | 3 | 2 | 3 | | 1 | 9 | 1 |

| | | | | |
|---|---|---|---|---|
| | | | Example 17 | |
| | | | Taking-off list | |
| | | | Power system | |
| | | | main switchboard | |
| | | | single sso | |
| | | | double sso | |
| | | | cooker cu | |
| | | | final circuits | |
| | | | RFI notes – Power | |
| | | | supply to switchboard? | |
| | | | distribution sheet? | |
| | | | cable containment info? | |
| | | | fire stopping? | |
| | | | testing & commissioning? | |
| | | | spec for terminal equip? | |
| | | | cable sizes? | |
| | | | builder's work? | |
| | 1 | 1 | Primary equipment, power system, location of installation low level on wall, main switchboard 39.1.1.1.1 | In this example, it has been assumed that the supply to the main switchboard and all associated work has been measured elsewhere. |

| | | | | |
|---|---|---|---|---|
| 1 | 1 | Terminal equiment and fittings, power system, single switched socket outlet, location of installation low level on walls 39.2.1.1.1 | |
| 9 | 9 | Terminal equiment and fittings, power system, double switched socket outlet, location of installation low level on walls 39.2.1.1.1 | |
| 1 | 1 | Terminal equiment and fittings, power system, cooker control unit, location of installation low level on walls 39.2.1.1.1 | |
| 1 | 1 | Final circuits, 2.5 mm$^2$ twin and earth, 10 points, location of installation as specification 39.7.1.1.0 (circuit 4 | |
| 1 | 1 | Final circuits, 10 mm$^2$ twin and earth, 1 point, location of installation as specification 39.7.1.1.0 (circuit 5 | |
| Item | | Testing power system 39.15.0.0.0 | |
| Item | | Commissioning power system 39.15.0.0.0 | |

*(Continued)*

*(Continued)*

| | | | | |
|---|---|---|---|---|
| | | | Taking – off list<br>Lighting system<br>lights<br>final circuits<br>switches | |
| | | | RFI notes – Lighting<br>full light spec?<br>installation locations?<br>cable spec?<br>testing & commissioning?<br>fire stopping details?<br>builder's work? | |
| | 10 | 10 | Terminal equipment and fittings, lighting system, recessed LED downlighter, location of installation high level on ceilings<br>39.2.1.1.1 | |
| | 1 | 1 | Terminal equipment and fittings, lighting system, feature pendant, location of installation high level on ceilings<br>39.2.1.1.1 | |
| | 3 | 3 | Terminal equipment and fittings, lighting system, external light, location of installation high level on walls, external work<br>39.2.1.1.1 | |
| | 2 | 2 | Terminal equipment and fittings, lighting system, 2–way light switch, location of installation low level on walls<br>39.2.1.1.1 | |

| | | | | |
|---|---|---|---|---|
| 3 | 3 | | Terminal equipment and fittings, lighting system, 1–way light switch, location of installation low level on walls<br>39.2.1.1.1 | |
| 2 | 2 | | Final circuits, 1 mm$^2$ twin and earth, 11 lighting points and 4 switch points, location of installation as specification<br>39.7.1.1.0<br>(circuits1 & 2 | |
| 1 | 1 | | Final circuits, 1 mm$^2$ twin and earth, 1 lighting point and 1 switch points, location of installation high level, external work<br>39.7.1.1.0<br>(circuit 3 | Note all work assumed internal unless described as otherwise. |
| Item | | | Testing lighting system<br>39.15.0.0.0 | |
| Item | | | Commissioning lighting system<br>39.15.0.0.0 | |

# 18

# Drainage Below Ground

## Subdivision

The measurement of drainage follows Work Section 34 and may be divided into the following sections:

- Manholes or inspection chambers.
- Main drain runs and fittings between manholes.
- Branch drain runs and fittings between outlets and manholes.
- Accessories (e.g. gullies).
- Sewer connection.
- Land drains.
- Testing.

If the foul water and the surface water drainage are separate systems, then usually each would be measured independently using the above sequence. When measuring surface water drainage, there must be liaison with the person measuring the roads and pavings in order to ascertain the position of gullies and channels. Similarly, there will have to be consultation with the person measuring the roofs in order to ascertain the position of rainwater pipes and to decide who will measure the connections if required. When measuring foul water drainage, there will have to be similar discussion with the person measuring the plumbing in order to find out the position of soil pipes and outlets and again to establish who is to measure the connections. The opportunity should be taken during these discussions to check that the drawings show drain runs leading from all service outlets.

## Manholes

In most cases, the position of manholes will be shown on the drawings together with invert levels. If the existing and finished ground levels adjacent to the manhole are not shown, then these will have to be ascertained from the site plans, possibly by interpolation.

*Willis's Elements of Quantity Surveying*, Fourteenth Edition. Roy Hills and Sandra Lee.
© 2024 John Wiley & Sons Ltd. Published 2024 by John Wiley & Sons Ltd.

| Manhole schedule | | | | | |
|---|---|---|---|---|---|
| Nr | Diagram | Intl size | Depth to invert | Conc base | Cover slab |
| 1 | | 900 × 570 | 900 | 150th | 900  570<br>430  430<br>1 330 × 1 000 |
| 2 | | 900 × 570 | 900 | 150th | 900  570<br>430  430<br>1 330 × 1 000 |
| 3 | | 900 × 570 | 750 | 150th | 900  570<br>430  430<br>1 330 × 1 000 |

**Fig. 39**

Prior to the measurement of manholes, it is wise to prepare a schedule of similar format to that shown in Figure 39. This relates to the manholes measured in Example 15.
If information on the sizing of manholes is not available, then the following guide may be useful:

- For inspection chambers up to 900 mm deep, the minimum internal size should be 700 mm wide × 750 mm long, allowing up to two branches on each side.
- For manholes over 900 mm and up to 3300 mm deep, the minimum size should be 750 mm wide × 1200 mm long, allowing 300 mm in the length for each 100 mm branch and 375 mm for each 150 mm branch.
- For manholes up to 2700 mm deep, a cover of size 600 × 600 mm should be provided, increasing to 600 × 900 mm for deeper manholes.
- For deep manholes, an access shaft can be constructed at the top to within 2 m of the benching.
- When the depth of the manhole exceeds 900 mm, step irons should be taken at 300 mm intervals.

Manholes, inspection chambers, and the like are enumerated with a detailed description of the construction and internal dimensions, giving the depth from the cover level to the invert level in 250 mm stages.

## Drain runs

Once the manholes have been measured, it is a comparatively simple matter to measure the main drain runs between. First, a schedule should be prepared on the lines of that shown in Figure 41. This schedule has been prepared for the drain runs in Example 15, and it could have replaced many of the waste calculations. The depth of the drain excavation will usually be the same as that for the manhole at each end, although a small adjustment may have to be made if the bed below the manhole is of different thickness to that below the drainpipe. The depths at each end of the drain run will be averaged, and the depth classified in 500 mm stages. Drain runs including the pipes are measured as linear items stating the type and nominal size of the pipe. If multiple pipes are included in the trench, then this should be stated. Earthwork support, treating bottoms, filling, and disposal are all deemed to be included with the trench item.

Bedding and surrounds for pipes should be given in the description of the drain run pipes.

The pipe length through the manhole wall, building in the end of the pipe, and any bedding or pointing are deemed included. Pipe fittings such as bends are enumerated, and the method of jointing given.

Branch drain runs are measured in the same way as main drain runs, and again a schedule should be prepared before measurement. The important difference is that branch runs are connected to the manhole at one end only. At the other end will be either a gully or a connection to a soil pipe or similar outlet. The depth of the trench at this end will depend upon the size of the gully or rest bend, but for normal circumstances, it could be taken as 600 mm. If bends are used, then it must be remembered that the run must be measured to include their length. Gullies and other accessories are enumerated and described, jointing to pipes and concrete bedding being deemed included. Back and side inlets, raising pieces, and gratings should be included in the description of the gully. Usually, two bends are required in a pipe coming from a gully as there may have to be a change in direction as well as an adjustment to achieve the correct fall. Circular-top or two-piece gullies may, however, reduce the need for direction change.

The connection to the public sewer, if carried out by the contractor, is enumerated and described. If the connection is carried out by the statutory authority, as is usually the case, then a provisional sum is included to cover the cost of the work. Care must be taken to ascertain the extent of the work done by the authority and to include for the remainder in the measured work. The work frequently involves excavation across the highway, which would require special provisions such as traffic control, lighting, and reinstatement. It should also be noted that work beyond the boundary of the site has to be kept separate.

Testing the drainage system is included as an item, stating the method to be used. The extent of testing required should be ascertained from the appropriate authority.

Trench excavation depth allows for 100 mm bed and 20 mm thickness of the pipe. The length is measured between outside face of manholes. Formation level has been assumed at 100.00, around house, and then falling away along the main runs.

Pipes are assumed to fall at least 1:40 (therefore over 30 m the drop is 750 mm and over 10 m the drop is 400 mm).

Pipes measured externally to house and up to manhole 1.

| Run location | Pipe size 100 mm | Form. level at head | Invert level at head | Exc. depth at head | Form. level at foot | Invert level at foot | Exc. depth at foot | Ave. depth of exc. | Depth in 500 mm increments | |
|---|---|---|---|---|---|---|---|---|---|---|
| | | | | | | | | | 0.50–1.00 m | 1.00–1.50 m |
| SVP-MH3 | 2.10 | 100.00 | 99.55 | 0.57 | 100.00 | 99.25 | 0.87 | 0.72 | 2.10 | |
| Gulley-MH3 | 1.00 | 100.00 | 99.55 | 0.57 | 100.00 | 99.25 | 0.87 | 0.72 | 1.00 | |
| WC to MH3 | 1.60 | 100.00 | 99.55 | 0.57 | 100.00 | 99.25 | 0.87 | 0.72 | 1.60 | |
| MH3-MH2 | 10.00 | 100.00 | 99.25 | 0.87 | 99.75 | 98.85 | 1.02 | 0.945 | 10.00 | |
| MH2-MH1 | 30.00 | 99.75 | 98.85 | 1.02 | 99.00 | 98.10 | 1.02 | 1.02 | | 30.00 |
| | | | | | | | | | | |
| PIPES | 44.70 | | | | | | | Length of trenches | 14.70 | 30.00 |

Fig. 41

Head is the shallow end of a run, foot is the deeper end of a run. Cover level (CL) is also commonly used instead of formation level.

To calculate excavation depth at head or foot is (formation level less invert) plus thickness of pipe and bed (120 mm).

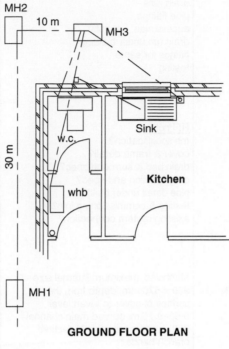

**GROUND FLOOR PLAN**

**Fig. 40**

| | | | | |
|---|---|---|---|---|
| | | | Example 15 | |
| | | | Taking – off list<br>manholes<br>covers and frames<br>drain runs<br>pipe fittings<br>accessories, i.e. gully<br>drain run under building<br>fittings for same<br>testing | |
| | | | RFI notes<br>mh specification?<br>cover & frame detail?<br>drain bed & surround spec?<br>gully location and spec?<br>pipe detail under building?<br>testing & commissioning info?<br>existing system connection? | |
| | 1 | 1 | Manholes, maximum internal size<br>900 × 570 mm, depth from the top<br>surface of cover to invert level<br>0.50–0.75 m, curved main channel<br>900 mm long with 2nr ¾ section<br>branch bends<br>34.6.2.6.0<br>(mh3 | |

| | | | |
|---|---|---|---|
| 1 | 1 | Manholes, maximum internal size 900 × 570 mm, depth from the top surface of cover to invert level 0.75–1.00 m, curved main channel 900 mm long 34.6.2.6.0 (mh2 | |
| 1 | 1 | Manholes, maximum internal size 900 × 570 mm, depth from the top surface of cover to invert level 0.75–1.00 m, straight main channel 900 mm long 34.6.2.6.0 (mh1 | |
| 3 | 3 | Covers and frames, 600 × 450 mm, bedded in cement mortar 34.14.1.1.1 | |
| 2.10 | 2.10 | Drain runs, average trench depth 0.50–1.00m, vitrified clay pipe 100 mm diameter, in situ concrete pipe bed and surround 34.1.1.2.0 | |
| 1.00 | 1.00 | | |
| 1.60 | 1.60 | | |
| 10.00 | 10.00 | | |
| | 14.70 | | |

*(Continued)*

*(Continued)*

| | | | |
|---|---|---|---|
| | 30.00 | 30.00 | Drain runs, average trench depth 1.00–1.50 m, vitrified clay pipe 100 mm diameter, in situ concrete pipe bed and surround 34.1.1.2.0 |
| 3/ | 2 | 6 | Pipe fittings, 100 mm diameter vitrified clay bend 34.3.1.0.0 |
| | 1 | 1 | Accessories, clay trapped gully as detail 34.4.1.0.0 |
| | 1.40 | 1.40 | Drain runs, average trench depth 0.50–0.50 m, cast iron pipe with spigot and socket joint 50 mm diameter, in situ concrete pipe bed and surround 34.1.1.2.0 (under bldg |
| | 2.20 | 2.20 | |
| | | 3.60 | |
| 2/ | 2 | 4 | Pipe fittings, 50 mm diameter cast iron bends 34.3.1.0.0 |
| | Item | | Testing and commissioning 34.17.0.0.0 |
| | Item | | Connection to existing manhole 34.16.1.0.0 |

# 19

# External Works

## Particulars of the site

Before the measurement of external works is commenced, a visit to the site should be made to ascertain items to be included or to check the information shown on the drawings. Items which could be checked or taken include:

- Grid of levels and dimensions of the site.
- Pavings, etc. to be broken up.
- Demolition of walls, fences, buildings, etc.
- Felling trees and grubbing up hedges.
- Preservation of trees and grassed areas, etc.
- Any existing services, overhead power lines, etc.
- Any turf that may be worth preserving.

In addition to noting these items, consideration could be given at the same time to other matters, such as access to the site, which may have to be drawn to the attention of the tenderers in the preliminaries.

## Coverage

The measurement of external works can include several different aspects of work. To give an idea of the coverage, a selection of items is listed below:

1) Site clearance including:
   - Removal of trees, hedges, and undergrowth
   - Lifting turf for reuse
   - Breaking up pavings, etc.
   - Demolition work.
2) Temporary works (other than those for the contractor's own convenience) including:
   - Roads
   - Fencing
   - Maintaining existing roads.

*Willis's Elements of Quantity Surveying,* Fourteenth Edition. Roy Hills and Sandra Lee.
© 2024 John Wiley & Sons Ltd. Published 2024 by John Wiley & Sons Ltd.

3) Roads, car parks, paths, and paved areas including:
   - Preparatory work
   - Pavings
   - Kerbs, edgings, channels
   - Drop kerbs
   - Steps and ramps
   - Road markings
   - Surface drainage.

4) Fencing and walls including:
   - Boundary, screen, retaining walls
   - Fencing, gates
   - Guard rails
   - In situ planters.

5) Outbuildings including:
   - Garages
   - Substations
   - Gatekeepers' offices
   - Bus shelters
   - Canopies.

6) Sundry furniture including:
   - Bollards
   - Seats and tables
   - Litter, grit, and refuse bins
   - Cycle stands
   - Prefabricated planters
   - Flag poles
   - Clothes driers
   - Sculptures
   - Signs and notices
   - Lighting standards.

7) Water features including:
   - Lakes and ponds
   - Ornamental and swimming pools
   - Fountains.

8) Horticultural work including:
   - Cultivating, topsoil filling, subsoil drainage
   - Grassed areas (seeding and turfing)
   - Planting trees (tree grids and guards)
   - Planting shrubs, hedges, and herbaceous plants.

9) Sports facilities including:
   - Playing fields and running tracks
   - Tennis courts
   - Bowling greens
   - Children's play equipment.

10) Maintenance including:
    - Planted and grassed areas
    - Playing fields.
11) External services including:
    - Water, gas, and electric mains
    - Telephone and TV services
    - Fire and heating mains
    - Security systems.

Whilst the above areas of work have been classified under various headings, the presentation of the work in the bill is a matter for personal preference. NRM2 provides specific measurement rules for roads and pavings, edgings, soft landscaping, fencing, and site furniture, and it would be as well to follow these at least.

## Site preparation

Under Work Section 5, trees and tree stumps to be removed from the site are taken as enumerated items, stating the girth measured 1.00 m above original ground level as 500 mm–1.50 m, 1.5–3 m, and over 3 m in 1.5 m stages. Grubbing up roots, disposal, and filling voids are deemed to be included although a description of the filling has to be given. Site clearance is measured as metres square and includes for all trees up to 500 mm girth. Lifting turf for preservation is given in square metres as is the removal of topsoil (depth stated) and the breaking up of existing pavings (type of paving and depth stated).

## Excavation

General excavation in connection with external works is measured under the same rules as for general building work. The main problems in the measurement of external works are likely to arise from bulk excavating to reduce levels in uneven ground and from measuring irregular areas. Guidance on calculations for these items is given in the Appendix. Provided that a proper grid of levels has been taken over the site, it should be fairly simple to find the average depth of excavation.

There may be a difficulty in giving the required depth stages for excavation when depths vary over the site. If this appears to be a problem and would mean drawing many contours representing depth changes, it is suggested that, irrespective of the required depth stages, an average depth is found for the area and this is used for the classification. The contractor should, of course, be informed that this method has been adopted. Frequently, it is prudent to divide large irregular areas into sections for measurement; not only will this assist with the calculation of the area but it will also give greater accuracy in the depth measurements and classifications. Furthermore, it may be wise to make a division at the edge of an area of shallow excavation where it becomes deeper. Handling excavated material is generally the responsibility of the contractor, but if there are any specified conditions regarding handling, such as the provision of temporary spoil heaps, then these have to be stated.

## Roads and paving

Specifications for roadworks frequently refer to guides and recommendations published by the local authority or government department. Concrete roads are measured as cubic items and classified as for slabs. Mechanical treatment to the surface of the concrete is given in square metres. Macadam roads are measured in square metres, the area taken being that in contact with the base or lineally if less than 300 mm wide with the thickness stated. No deductions are made for voids within the area when not exceeding $1.00\,m^2$. Work is described as being level and to falls only, to falls $\leq 15°$ and $> 15°$ from horizontal. Forming or working into shallow channels is included but linings to channels are taken as linear items, stating the thickness and girth on face; all labours are deemed included. Gravel paving and roads are measured and classified in the same way as macadam.

Brick and block paving is also measured in a similar manner. Sports surfacing is also measured under the rules in Work Section 35. Kerbs, edgings, and channels are measured linear, with specials (such as corner blocks or outlets) being enumerated as extra over items.

## Walling

Walling in connection with external works is again measured in accordance with the general rules although particular attention needs to be given to the measurement of any curved and battered work.

## Fencing

Fencing is measured under Work Section 36 as a linear item over the posts. There are several specifications for fences and these can frequently be used to assist with descriptions. Posts or supports occurring at regular intervals are included in the description of the fencing but occasional supports such as straining posts are enumerated and described as extra over the fencing. The excavation of post holes, backfilling, earthwork support, and disposal of surplus are deemed included. Gates are enumerated and ironmongery is also enumerated separately.

## Sundry furniture

Most of the items listed above under 'Coverage', item (6), Sundry furniture, are enumerated and supported by a component drawing, dimensioned diagram, or reference to a trade catalogue or standard specification.

## External services

As far as services are concerned under Work Section 41, pipes in trenches have to be kept separately. Special rules apply to the measurement of trenches for services; these are measured as linear items stating the nominal size and type of service. Average depths are stated in 500 mm stages; earthwork support, backfilling, and disposal are deemed included.

Example 16 is of a simple estate road and path and the associated bell-mouth junction with a main road.

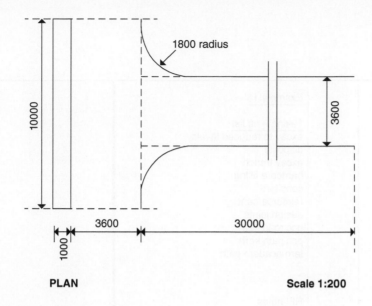

1800 radius

10000

3600

3600

1000

3600

30000

**PLAN**

**Scale 1:200**

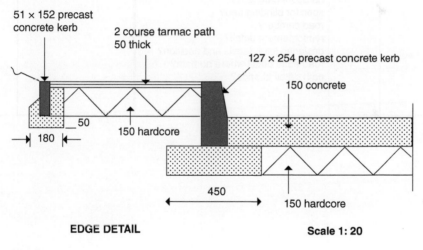

51 × 152 precast concrete kerb

2 course tarmac path 50 thick

127 × 254 precast concrete kerb

150 concrete

50

150 hardcore

180

450

150 hardcore

**EDGE DETAIL**

**Scale 1: 20**

**Fig. 42**

**Example 16**

**Roads and Paths**

| | | | | |
|---|---|---|---|---|
| | | | Example 16<br><br>Taking – off list<br>excav to reduced levels<br>disposal<br>excav trench<br>hardcore filling<br>conc bed<br>reinforcement<br>design joints<br>pcc road kerb<br>pcc path kerb<br>tarmacadam path<br><br><br>RFI notes<br>chk excav starting level<br>geotextile mat for road & path?<br>hardcore camber/fall?<br>spec for blinding bed?<br>road camber?<br>reinforcement detail?<br>designed joint details and position?<br>road edge detail where no path?<br>path detail at ends? | |

| | | | | |
|---|---|---|---|---|
| | | | Excavation, starting level topsoil reduced level, bulk excavation, not exceeding 2 m deep 5.6.1.1.0 | |
| 2/ 3/14/ | 40.00 4.18 0.40 | 67.50 | (road | road lgth 30.000 10.000 40.000 |
| | 1.80 1.80 0.40 | 0.56 | (road bellmouths | Area of a bellmouth is $3/14r^2$ |
| | | | | road kerb spread fdn 0.450 ddt road kerb 0.127 0.5/ 0.323 0.162 |
| | 10.00 0.83 0.20 | 1.65 | (path | road width 3.600 kerb b.s. 2/ 0.127 0.254 3.854 spread b.s. 2/ 0.162 0.323 4.177 |
| | 10.00 0.18 0.05 | 0.09 69.81 | (path kerb fdn | rd depth kerb 0.254 hardcore 0.150 0.404 |
| | | | | path kerb spread 0.180 ddt kerb 0.051 0.5/ 0.129 0.065 |

*(Continued)*

(Continued)

| | | | |
|---|---|---|---|
| | | | path width |
| | | | 1.000 |
| | | | ddt road kerb 0.127 |
| | | | 0.873 |
| | | | ddt spread 0.162 |
| | | | 0.712 |
| | | | path kerb 0.051 |
| | | | 0.763 |
| | | | spread 0.065 |
| | | | 0.827 |
| | | | |
| | | | path depth |
| | | | path 0.050 |
| | | | hardcore 0.150 |
| | | | 0.200 |
| | | | |
| | | | & |
| | | | Disposal, excavated material off site |
| | | | 5.9.2.0.0 |
| | | | |
| | | | Imported filling, beds, and voids over 50 mm thick but not exceeding 500 mm thick, finished thickness 150 mm, level |
| | | | 5.12.2.1.0 |
| | 40.00 | | (road |
| | 3.28 | | road width |
| | 0.15 | 19.66 | 3.600 |
| | | | ddt spread b.s.2/0.162 0.323 |
| 2/ | | | 3.277 |
| 3/14/ | 1.80 | | (bellmouths |
| | 1.80 | | |
| | 0.15 | 0.21 | path width |
| | | | 1.000 |
| | 10.00 | | (path |
| | 0.87 | | ddt road kerb 0.127 |
| | 0.15 | 1.31 | 0.873 |
| | | 21.18 | |

Formula for the area of a segment is $3/14r^2$.

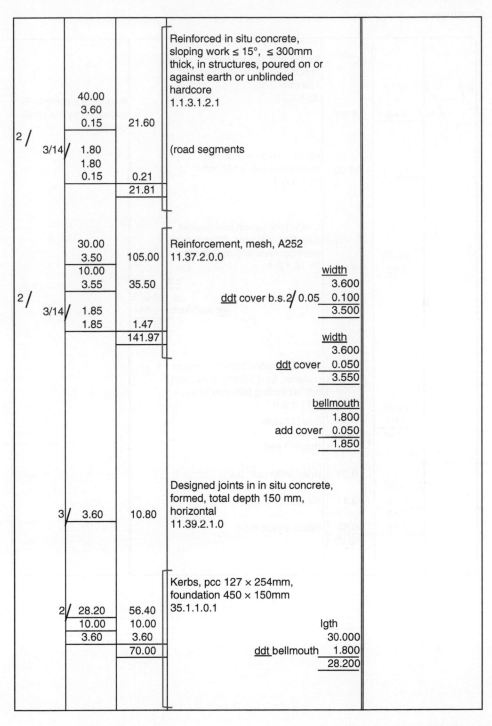

| | | | |
|---|---|---|---|
| | 40.00 | | Reinforced in situ concrete, sloping work ≤ 15°, ≤ 300mm thick, in structures, poured on or against earth or unblinded hardcore |
| | 3.60 | | |
| | 0.15 | 21.60 | 1.1.3.1.2.1 |
| 2/ 3/14/ | 1.80 | | (road segments |
| | 1.80 | | |
| | 0.15 | 0.21 | |
| | | 21.81 | |

| | | | |
|---|---|---|---|
| | 30.00 | | Reinforcement, mesh, A252 |
| | 3.50 | 105.00 | 11.37.2.0.0 |
| | 10.00 | | |
| | 3.55 | 35.50 | width 3.600 |
| 2/ 3/14/ | 1.85 | | ddt cover b.s.2/ 0.05   0.100 |
| | 1.85 | 1.47 | 3.500 |
| | | 141.97 | width 3.600 |
| | | | ddt cover   0.050 |
| | | | 3.550 |

bellmouth
1.800
add cover  0.050
1.850

| | | | |
|---|---|---|---|
| | | | Designed joints in in situ concrete, formed, total depth 150 mm, horizontal |
| 3/ | 3.60 | 10.80 | 11.39.2.1.0 |

| | | | |
|---|---|---|---|
| | | | Kerbs, pcc 127 × 254mm, foundation 450 × 150mm |
| 2/ | 28.20 | 56.40 | 35.1.1.0.1 |
| | 10.00 | 10.00 | lgth 30.000 |
| | 3.60 | 3.60 | ddt bellmouth   1.800 |
| | | 70.00 | 28.200 |

(*Continued*)

*(Continued)*

| | | | | |
|---|---|---|---|---|
| $\dfrac{2}{1/4}\Big/$ 3.14/ | 3.60 | 5.66 | Kerbs, pcc 127 × 254 mm, curved, radius 1.80 m, foundation 450 × 150 mm<br>35.1.1.01. | Radius of a circle is Pi multiplied by the circle diameter. |
| | 10.00 | 10.00 | Kerbs, pcc 51 × 152 mm, foundation 450 × 150 mm<br>35.1.1.0.1 | |
| | 10.00<br>0.82 | 8.22 | Coated macadam and asphalt, over 300 mm wide, 50 mm thick, pavings, level and to falls only<br>35.12.1.2.1 | |

           path width

         ab   0.873

ddt path kerb  0.051

          0.822

| | | | |
|---|---|---|---|
| 10.00<br>0.06 | 0.65 | Filling obtained from excavated material, final thickness of filling not exceeding 500 mm deep<br>5.11.1.2.0<br>(rear of path | |
| 10.00<br>0.16 | 1.62 | (front of path | |
| 5.66<br>0.16 | 0.91 | (road bellmouth b.s | |
| 2/ 28.20<br>0.16 | 9.11 | (long side of road | |
| 3.60<br>0.16 | 0.58 | (short end of road | |
| | 12.86 | | |

| | | | |
|---|---|---|---|
| | | | ddt |
| | | | disp excav mat offsite abd |
| | 10.00 | | (rear of path |
| | 0.20 | | |
| | 0.15 | 0.30 | |
| | 10.00 | | (front of path |
| | 0.18 | | |
| | 0.15 | 0.27 | |
| | 5.66 | | (road bellmouth b.s. |
| | 0.16 | | |
| | 0.25 | 0.23 | |
| 2/ | 28.20 | | (long side of road |
| | 0.16 | | |
| | 0.25 | 2.31 | |
| | 3.60 | | (short end of road |
| | 0.16 | | |
| | 0.25 | 0.15 | |
| | | 3.27 | |

# 20
# Preliminaries and Other Priced Bill Sections

## Generally

The drafting of the preliminaries, risks, provisional sums, works to be carried out by statutory undertakers, and dayworks sections of the bill, which are not usually derived from the dimensions, is discussed in this chapter. These sections are often left to a late stage in the bill preparation, as reference to the draft bill of measured items may be necessary. The specification will form the basis of these sections; if items have been lined through in the specification as they are measured, then those remaining may have to be either included or, if the specification is being provided with the tender documents, referred to in these parts of the bill. It should be noted, however, that specification items that relate to method and quality of workmanship are usually excluded from the bill unless they affect cost.

Often, companies will use standard preliminaries as the basis of creating the preliminaries bill to reduce drafting time and force the quantity surveyor to continually question the need for particular items. However, care needs to be taken to ensure that specific items required for the project are not overlooked.

## Preliminaries and general conditions section

It must be remembered that the bill of quantities has to set out all the circumstances and conditions that may affect the contractor's tender. Items that are of a general nature and do not necessarily relate to the quantity of permanent work are set out in the preliminaries section for pricing by the estimator. Work Section 1 of NRM2 comprises the rules for describing and quantifying preliminaries. It is not possible to quantify the main contractor's preliminaries in the way that a work section is quantified as it is the contractor's responsibility to interpret the information provided in the tender documents and to make necessary allowances for his or her chosen method of working and the resources required to complete the project. For this reason, the preliminaries bill usually looks more like a list of headings to be priced. The section is divided as follows:

- Part A: Information and requirements (i.e. dealing with the descriptive part of the preliminaries).

*Willis's Elements of Quantity Surveying*, Fourteenth Edition. Roy Hills and Sandra Lee.
© 2024 John Wiley & Sons Ltd. Published 2024 by John Wiley & Sons Ltd.

- Part B: Pricing schedule (i.e. providing the basis of a pricing document for preliminaries).

Both Part A and Part B are split to cater for whether the bill is for the main contract works or a works package. There are differences between these sections; however, they cover similar areas, and for clarity only the main works portion is covered here.

### Part A: Information and requirements

It should be noted that although items are listed below as they are shown in NRM2, they may also be specifically referred to in the construction contract for the works.

*Project particulars (A.1)*
This section sets the scene as to the proposed contract and full particulars of the project, including the location and details of the employer and consultants.

*Specifications and drawings (A.2)*
The bill should contain a list of specifications and drawings from which the bill was prepared and other documents related to the contract but not included in the tender documents. There should also be a statement explaining how the preconstruction information is dealt with in the tender documents.

At the commencement of each work section in NRM2, the exact nature of the drawn information to be given is set down. Three types of drawings have to accompany the bill of quantities:

- Block plan – to identify the site and outline of the building in relation to the town plan or other wider context.
- Site plan – to locate the position of the building in relation to setting out points, means of access, and general layout of the site.
- Plans, sections, and elevations – to show the position occupied by the various spaces in a building and the general construction and location of the principal elements.

*The site and existing buildings (A.3)*
A full description of the site and details of any existing buildings and services should be given if their presence is likely to affect the cost of the new work. Examples are a mediaeval listed building or a multi-storey car park with continuous traffic movements. Other details may include the results of the site investigation and the health and safety file.

*Description of the works (A4)*
This includes information on the project objectives to explain the context of the project. Details and dimensions of the new buildings and how they are to be constructed must be set out by the provision of either drawn information or measurement descriptions.

*General Constraints on executing the works (A.5)*
There are 23 listed conditions covering general constraints including working hours, deliveries, parking, etc.

*Contract conditions (A.6)*

Particulars of the contract have to be given, but if a standard published form of contract is being used, then the conditions need not be repeated in full in the bill provided that the form of contract together with any relevant amendments is named and the numbers and titles of the clauses are listed. The decision as to which clauses require pricing is left to the estimator. If the conditions of contract are not standard, then the conditions must be included in full in the bill or alternatively presented as a separate volume with the tender documents and listed in the bill for pricing as shown in this chapter. Special clauses may sometimes be added to the conditions of a standard form of contract, and these together with any modifications to clauses should be detailed in the bill. Such amendments must also be made to the actual contract that is to be signed by the employer and successful contractor, as it is not advisable to rely on a clause in the bill alone even though it may be one of the contract documents.

*Employer's requirements: provision, content, and use of documents (A.7)*

This section provides any applicable definitions and interpretations. Documents provided by different parties are also listed.

*Employer's requirements: contractor-designed work (A.8)*

In this section, the design responsibility is included along with information on how the contractor's design may be used.

*Employer's requirements: building information modelling (A.9)*

Describes the information model requirements and specifies the procedures to be followed including the common data environment requirements.

*Employer's requirements: completion (A.10)*

This section covers completion, sectional completion, and early possession requirements, if known. There is also an opportunity to include training of building maintenance staff and any specific requirements for operation and maintenance manuals.

*Employer's requirements: programme/progress (A.11)*

In this section, the format and content of programmes are described. A client may also have deadlines for programmes to be submitted so that the employer has time to assess the content before any planned meetings.

*Employer's requirements: management of the works (A.12)*

This includes any specific communication systems such as online collaborative tools and standard templates. Any particular photography requirements can also be described here.

*Employer's requirements: working with the employer and others (A.13)*

Shared working areas and information on coordination and cooperation are to be shown in this part. An example could include a Permit to Work system so that a contractor's work can be safely integrated into the employer's business as usual operations.

*Employer's requirements: quality management (A.14)*

This section allows the employer to state any requirements for the contractor's quality management system including the quality policy statement and quality plan. There may be a need for a contractor to align the timing of any quality audits with the employer's quality procedures. Mock-ups and samples could also be described here.

*Employer's requirements: tests and inspections (A.15)*

The management of tests and inspections are included here and can include the timing of any tests along with information on any tests that the employer's personnel might be required to observe. This also includes any statutory inspections such as those which may be required under the Building Regulations.

*Employer's requirements: services and facilities (A.16)*

Under this section, services and facilities provided by the contractor for the employer's use could include site accommodation such as an office and welfare facilities for the employer's sole use. Services and facilities provided by the employer for the contractor are also to be listed here.

*Employer's requirements: health and safety (A.17)*

This can include information on the employer's safety management procedures reporting process for the contractor. The format, type, and timing of method statements and risk assessments can also be listed.

*Employer's requirements: subcontracting (A.18)*

Details include any restrictions or specific requirements for the use of subcontractors.

*Employer's requirements: title (A.19)*

This refers to the requirements to ensure transfer of title to the employer such as storage requirements and marking. Any credit for salvaged materials may also be included.

*Employer's requirements: records (A.20)*

This section allows the employer to state the format and type of record required by the employer and may include weather conditions.

## Part B: Pricing schedule

In Part B, the pricing schedule tables list the components to be priced and identify what is to be included in the detailed description. The pricing tables give the unit of measure to be used for each item and require the items to be classified as either a 'fixed charge' or a 'time-related charge'.

A *time-related charge* is one that is considered to be proportional not to the quantity of the item but to the length of time taken to execute the work. A *fixed charge* is one that is considered to be proportional to neither the quantity of work nor the time taken.

Broadly speaking, this part of the preliminaries section contains two types of items: first, those that have a separate identifiable cost such as insurance, site facilities, and various fees payable; second, those costs that arise from a particular method of carrying out the work. The latter would include fixed costs for items such as providing the plant and bringing it to and removing it from the site, and time-related costs for items such as maintaining the plant on-site or for providing supervision for the works.

*Employer's requirements: site accommodation (B.1)*
Employers' requirements include the provision of site offices and any associated furniture and equipment, fences, and name boards. The delivery and installation, and subsequent removal, would be considered as fixed charges. The hire and maintenance, i.e. cleaning, would be assessed as time-related on a weekly basis.

*Employer's requirements: site records (B.2)*
The site records and costs of any web-based information management systems are shown as a fixed charge.

*Employer's requirements: completion and post-completion requirements (B.3)*
Elements in this section may consist of fixed charges such as the provision of spare parts for a predetermined period or time-related for the operation and maintenance of specified building engineering services installations.

*Contractor's cost items: management and staff (B.4)*
Any project-specific management and staff are assessed on a time-related basis. Depending on the nature of the project, there may also be costs with visits to subcontractors' works or consultants' offices.

*Contractor's cost items: site establishment (B.5)*
Similar to the employers' requirements for site accommodation, there are likely to be a mixture of fixed and time-related costs.

*Contractor's cost items: temporary services (B.6)*
This section covers costs associated with the provision of any temporary service and is also likely to be a mixture of fixed costs, i.e. temporary connections, and time-related such as hire for equipment.

*Contractor's cost items: security (B.7)*
Security staff are logically assessed as a time-related cost but there may also be security equipment, i.e. hoardings and gates which have both fixed costs for the installation and removal and also time-related for the hire of those elements.

*Contractor's cost items: safety and environmental (B.8)*
Similar to previous sections, equipment for safety and environmental protection may have both fixed and time-related costs. For example, scaffolding costs could consist of erection/ adaption/removal as fixed costs and the hire element as time-related.

*Contractor's cost items: control and protection (B.9)*
Surveys such as environmental and topographical may be a fixed charge particularly if they are carried out by specialist consultants/subcontractors. However, if monitoring is required, then this would logically be assessed on a time-related basis.

*Contractor's cost items: mechanical plant (B.10)*
Large items of plant are likely to have fixed costs, i.e. delivery and/or installation charges for the erection of cranes. Once assembled, there would be hire charges which would be captured as time-related charges. When no longer required, the dismantling and removal from site would be a fixed charge.

*Contractor's cost items: temporary works (B.11)*
This component includes temporary works for common use on site and may include, for example, access scaffolding which would therefore be assessed under the headings of fixed and time-related costs.

*Contractor's cost items: management and staff (B.12)*
Site records such as photography and time lapse videos have both fixed and time-related charges, i.e. the former to setup/dismantle and the latter to maintain the equipment.

*Contractor's cost items: completion and post-completion requirements (B.13)*
The cost of preparing of any commissioning plan or handover plan is assessed under this component heading.

*Contractor's cost items: cleaning (B.14)*
Elements under this component heading include maintenance such as keeping the site clean and also the final builder's clean.

*Contractor's cost items: fees and charges (B.15)*
Examples of charges under this component included oversailing fees which might arise if cranes are being used. Other charges might relate to the suspension of parking bays.

*Contractor's cost items: site services (B.16)*
As an example, this section might include a multi-service gang whose cost is not directly allocatable to a specific measured works element.

*Contractor's cost items: insurance, bonds, guarantees, and warranties (B.17)*
Charges associated with these components may be allocated here if they are not part of the overheads cost. Project-specific requirements should be included here.

The following describes sections which may also be found in bills of quantities and refers to non-measured works. These are fully described in NRM2 under 2.5.5.

*Provisional sums (2.5.5)*

These are included for work for which there is insufficient information available for proper description. Provisional sums may also be included to cover possible expenditure on items that may be required but for which there is no information available at tender stage. Here, NRM2 states that provisional sums may be for either undefined or defined work (i.e. work that can be described fairly fully). For the latter, the contractor is expected to have taken the work into account when pricing other sections, such as indirect costs in preliminaries. The effect of this is that when the actual sum expended is ascertained for the final account, no adjustment to other prices in the bill is made. On the other hand, if the work is undefined, then preliminaries such as, for example, plant items may have to be adjusted when the work is valued for the final account.

All provisional sums shall be exclusive of overheads and profit.

*Works by statutory undertakings (NRM2 Part 2–2.9.4)*

Works that are required to be carried out by a statutory undertaker are included as a 'provisional sum' under this separate section. The scope of the works is described, and the contractor is deemed to have made allowance for any time and cost implications of the works, including all general attendances.

The provisional sums for work to be carried out by statutory undertakers do not include an amount for overheads and profit. The bill includes a separate section for this.

*Risks (2.5.6)*

At the time of preparing a bill of quantities, there will normally still be a number of risks remaining to be managed by the employer and the project team. It may be considered appropriate for some of these to be managed by the contractor. A schedule of construction risks should be included in the bill of quantities for those risks that the employer wishes to transfer to the contractor.

The price inserted against these risks (effectively, the contractor's estimation of the time and cost implications should the risk arise) is to be paid, irrespective of whether the risk occurs.

In order for the contractor to effectively price the risks, they should be fully described so that it is transparent what risk the contractor is required to manage, and precisely what the employer is paying for. The allowances for construction risks do not include an amount for overheads and profit. These are included in a separate section. How the risks are managed will be decided based on the employer's risk management strategy and may also include sharing the risk with the contractor or retaining the risk by the employer.

*Dayworks (2.5.9)*

Daywork is work for which the contractor is paid on the basis of the cost of labour, materials, and plant plus an agreed percentage for overheads and profit. Payment in this way is usually reserved for items that cannot be measured and priced in the normal way. Daywork payments may arise in contract variations for unique items such as breaking up unexpected obstructions in the excavations or for adjustment of provisional sums. To enable a percentage rate for overheads and profit to be established at the tender stage, it is necessary to include items for these in relation to dayworks in the bill and, in order to obtain competitive rates, these items should be included in such a way that they become part of the tender sum.

One method of achieving this is to include a number of hours for both labourers and craftsmen which the estimator prices at an hourly rate and extends into the cash columns of the bill. Some surveyors split these hours into various crafts; of course, it needs to be made clear that these rates must include for all charges in connection with the employment of labour, plus profit and overheads, and so on. Alternatively, provisional sums can also be included in the bill for labour, plant, and materials, and the tenderer is invited to add a percentage that represents what will be added to the net cost of these items for incidental costs, overheads, and profit, should the need for daywork arise. This alternative may also refer to standard schedules such as the definition of prime cost of daywork carried out under a building contract.

# 21

# Bill Preparation

As described in Chapter 5, the widespread utilisation of computerised systems has to a great extent made the labour-intensive manual processing of dimensions redundant, and therefore only brief mention of such processes is made here. Preparation of the bill itself is addressed in more detail because (although it too is now often part of an automatic process), as with setting down dimensions, it is important to understand the process involved in structuring and coding of the bill; thus, when computerised systems are used, the surveyor can ensure that the bill is complete and correct and that the necessary level of information is being provided to the contractor for pricing.

## Abstracting

In splitting up the building into its constituent parts for measurement, the taker-off follows a systematic method, but this then creates a repetition of the same description in different parts of the dimensions. For example, facing brickwork will be measured when the external wall is measured, and again it will be in the windows and doors element as it is deducted during the measure of the openings.

Under the traditional method of processing dimensions, the function of the abstract was to collect similar items together and to classify them into sections and subsequently, according to certain accepted rules of order and arrangement, to put them in a suitable sequence for writing the bill. A manual abstract was prepared by copying the descriptions and squared dimensions from the taking-off onto abstract paper in a tabulated form as nearly as practicable in the order of the bill.

The need for an abstract is eliminated by the use of a computerised system; however, the principles behind the abstract are retained, and a full breakdown of any item can be obtained.

## Procedure

Traditionally, the writing of the bill (*billing*) involved, in theory, copying out the descriptions and quantities from the abstract in the form of a schedule or list on paper ruled with

*Willis's Elements of Quantity Surveying*, Fourteenth Edition. Roy Hills and Sandra Lee.
© 2024 John Wiley & Sons Ltd. Published 2024 by John Wiley & Sons Ltd.

cash columns for pricing, but, in practice, it was a good deal more than this. The taker-off may describe an item briefly if it is in common use and leave it to the billing stage for the full description to be compiled. In fact, the term *working-up* originated from the task of working-up or expanding the brief taking-off descriptions into proper bill items. With the use of standard libraries of descriptions, as long as the correct description has been chosen, then this should not be required. However, there is a need for a final read-through of the bill by someone who understands the project details, including its specification, in order to provide a final edit and make sure that the correct items have been included.

## Division into sections

The structure of bills of quantities varies depending on the complexity of the project and work measured. In certain countries it is customary to invite separate tenders for each trade, but in the UK, unless a form of management contracting has been adopted, it is usual to invite tenders from a general contractor only. Where separate tenders are invited, the need for separate bills for each trade is obvious, but even where tenders are obtained from a general contractor, the division into work packages is of assistance in pricing and simplifies the estimator's task in respect of the parts of the work to be sublet. Moreover, it is this first step in the subdivision of the bill which enables items to be easily traced. The sections into which the bill is divided generally align with subcontractors' work, and any departures, such as precast concrete bollards being contained in furniture or equipment, are soon recognised.

Work to existing buildings, work outside the curtilage of the site, and work to be subsequently removed have to be billed in separate sections. Traditionally, substructure or work up to and including damp-proof course and external works are also kept in separate sections. These indications of location should assist the estimator in pricing and furthermore facilitate the valuation of, and measurement of possible amendments to, this work at a later stage. When a project consists of several buildings, it is often desirable to provide a separate section in the bill for each one.

## Structure of bills

There are three principal breakdown structures for bills. They are:

1) *Elemental*: Measurement and description are by group elements, following the arrangement for elemental cost planning as defined in NRM1. Each group element forms a separate section of the bill and is subdivided through the use of elements, which are further subdivided by sub-elements.
2) *Work section*: Measurement and description are divided into the work sections defined in NRM2. (See Chapter 7 on taking-off.)
3) *Work package*: Measurement and description are divided into employer or contractor-defined work packages. Works packages can be based on either a specific trade package or a single package comprising a number of different trades.

For certain reasons, it may be desirable to produce an elemental bill, in which the main divisions are design elements or constituent parts of the building (e.g. foundations or floor construction). The main purpose of such a bill is to assist a standardised system of cost analysis, which may be adopted particularly where buildings of a similar nature are to be repeated. Whilst in theory this type of bill should make estimating more accurate because the items are related to a particular part of the building, contractors who sublet work often have difficulty in collecting appropriate items together. Furthermore, estimators may find similar items occurring in different elements and have the additional problem of relating the prices for these separated items. To overcome these difficulties, some surveyors offer tenderers the bill in either traditional or elemental formats; computer sortation makes this particularly easy. If a bill has been coded correctly at take-off, then, by a simple computer sorting process, a trade bill priced by a tenderer can be easily turned into an elemental bill and compared with the pre-tender estimate, thus aiding cost management.

## General principles

It must be remembered that in most cases the bill is a contract document, and therefore the descriptions must be absolutely clear and usually without abbreviations, other than those in common use. The biller must have sufficient knowledge of construction to understand the descriptions; if any appear to be vague or ambiguous, the taker-off should be consulted as to the exact meaning and an amendment made if necessary.

## Order of items in the bill

The order of items in the bill is a matter for personal preference but the Appendix to NRM2 provides guidance on alternatives. However, the following general principles are normally applied:

1) Work sections as contained in NRM2, although locational sections such as substructure or external works may be required.
2) (a) Subdivisions of work sections as contained in NRM2.
   (b) Subdivisions of different types of materials such as different mixes of concrete or different types of brick.
3) Within each subdivision in stage (2), the order of cubic, square, linear, and enumerated items.
4) Labour-only items should precede labour and material items within the subdivisions in (3).
5) Items within each subdivision in (3) and (4) are placed in order of value, often with the least expensive item first.
6) Preambles and prime cost and provisional sums usually form a separate bill, although in certain circumstances they may be contained in the appropriate work section.

## Format of the bill

The bill for each work section should be commenced on a new sheet. The ruling of the paper and a typical heading are:

| | | | £ | | |
|---|---|---|---|---|---|
| | BRICK WALLING | | | | |
| | Common brickwork in cement, lime, and sand mortar (1:1:6) | | | | |
| A | Walls, half brick, vertical | 97 | m² | | |

   The first (left-hand) column is for the item number or letter references and the binding margin. Sometimes, an additional line is ruled to separate the binding margin from the reference column. The main wide column is for headings, subheadings, and descriptions. Frequently, as illustrated here, an unruled portion is left at the top of each page for main headings such as work sections. The next columns contain the quantity of the item and the unit of measurement. Some surveyors prefer to enter the unit of measurement first so that the quantity is adjacent to the rate in the next column to facilitate the extension of the cash sum; others prefer to see the measurement figure separated from the cash figure so that there can be no confusion between the two. The last three columns are left blank for use by the estimator, who enters the rate for the item per unit of measurement and the cash extension. Sometimes, however, a cash amount is entered by the surveyor in the last two columns, for example when entering provisional sums.

## Referencing items

It is essential that every item in the bill can be referred to and found easily. To this end, it is preferrable to use the page number followed by a letter entered alphabetically against each item on the page with each page commencing with the letter 'A'. The letters 'I' and 'O' are usually omitted to avoid any possible confusion with numerals.

## Units of measurement

Great care must be taken, particularly when there is a change, to enter the correct unit of measurement. For example, an item measured as a cubic item and indicated as a superficial item in the bill may result in a considerable difference in price. Even if the error is detected and queried by a tenderer, it may involve the issue of an addendum to the bill – which should be avoided if possible. To assist in avoiding mistakes, particularly when items are inserted, it is preferable to enter the unit of measurement against each item rather than using two dots to indicate repetition.

## Order of sizes

Sizes or dimensions in descriptions should always be in the following order: length, width, depth, or height. Sometimes the width or front-to-back dimension of, say, a cupboard is referred to as its 'depth'. If there is likely to be any doubt, the dimensions should be identified. For example:

sink base unit 1000 mm long × 600 mm wide × 900 mm high overall

## Use of headings

Generous use of headings will not only help estimators to find their way about the bill but also may reduce the length of descriptions. Headings generally fall into one of four categories:

1) Work section headings, such as 'Masonry'.
2) Subsection headings, such as 'Brick walling'.
3) Headings that partly describe a group of items to follow, such as 'Common brickwork in cement mortar (1:3)'.

Work section headings are often repeated in the top right-hand corner of each page. Other headings should be repeated when a new page is started. This enables it to be seen at once what is being dealt with, thus avoiding the need to search back a few pages. A new heading usually terminates a previous heading; where a new heading is not used, or if there is likely to be any doubt, 'end of . . . ' should be entered in the appropriate place in the bill.

## Writing short

It is often convenient for pricing purposes to keep items together in the bill which would, following the normal rules, be separated. For example, it is probably easier for an estimator to price the fittings on a gutter whilst dealing with the particular gutter although this means that enumerated items have to be dealt with in the middle of linear items. The method of entering these items is known as writing short; it avoids breaking completely the sequence of linear items, as follows:

| D | 112 mm UPVC straight half round gutter fixed to timber with standard brackets | 125 | m | | | |
|---|---|---|---|---|---|---|
| E | Extra for stop end | 9 | nr | | | |
| F | Extra for angle | 7 | nr | | | |

As the example shows, written short items are inset so that the main items stand out. In the written short items, the words 'extra for' have been repeated instead of using 'ditto', thus avoiding any possible confusion with the use of 'ditto' in the main item. Items that are written short automatically refer to the main item under which they are written, so there is no need to refer back. NRM2 may require items that are often written short to be described as 'extra over'; this means that the main item has been measured over the written short item. In other words, the estimator is to allow for the extra cost of the item only as it displaces the main item.

## Unit of billing

The general unit of billing for other than enumerated items is the metre. The dimensions and their collections having been taken to two decimal places, the bill entries are rounded up or down to the nearest metre. The main exceptions are steel bar reinforcement and structural steel, which are billed in tonnes to two decimal places.

## Framing of descriptions

The framing of descriptions was introduced in Chapter 4 and should follow the standard NRM2 phraseology. Care must be taken to leave no doubt as to their meaning, particularly as the bill is a legal document. The opening phrase of a description should indicate the principal part of the item so that the estimator knows immediately what it is about. The word 'approved' should be avoided if it leaves any doubt about the quality of the item. Its sole justification is to point out to the contractor that the architect's approval should be obtained for the article to be used where there are several alternatives all costing about the same. If some particular material is described and the words 'or other approved' are added to the description, a gambling element is at once introduced which is directly contrary to the purpose of the bill of quantities. One estimator may decide to price for a less expensive alternative in the hope that the architect will approve it, whilst another may price the material specified. The words 'or other equal and approved' are sometimes used by public authorities after the specification of a proprietary article to avoid the criticism that a monopoly situation has been created, favouring a particular firm.

Long-winded descriptions should also be avoided, and any superfluous words omitted. The surveyor must cultivate the difficult art of making descriptions concise and easily understood and at the same time omitting nothing that is essential to the estimator for pricing. In the following, for example, the words underlined could be left out:

Small 19×19 mm wrot softwood cover fillet planted on around bathroom door frame
50 × 175 mm sawn softwood in floor joists spiked to timber wall plates.

It should be noted that in the above, if fixing is not specified, it is at the discretion of the contractor. In these cases, planting or spiking would normally be omitted as it is the most economical way of fixing. Furthermore, the location of an item is not normally required.

Consistency in spelling is important if only to make the bill look like a workmanlike document. Often, there are alternatives for spelling technical words such as 'lintel' or 'lintol' and 'cill' or 'sill'. An easy solution to this problem is to establish a *house rule* that spelling should be as shown in NRM2. Furthermore, consistency in language is important, as minor differences in phraseology may cause an estimator trouble in deciding if the meaning is intended to be different. For example:

1) Raking cutting *to* existing Code 4 milled lead
2) Curved cutting *on* existing Code 4 milled lead.

In this case, there is no actual difference to be inferred from the word 'on'.

## Totalling pages

There are two ways in which the surveyor may indicate how the cash totals on each page are to be dealt with. First, the total may be carried over to be added to the next page and so on to the end of the section. The foot of the page is completed as follows:

Carried forward £

and the top of the following page as:

Brought forward £

The end of the section is completed as follows:

Total Carried to Summary on page 796 £

This method is suitable only when the section contains a few pages. Any mathematical error made at an early stage is carried through all the pages and, when discovered, involves several corrections and re-totalling of the pages. Generally, this method should be avoided.

Second, each bill page may be totalled and each of these totals carried to a collection at the end of the section. The bottom of each page is completed as follows:

Carried to Collection on page 84 £

The collection is on the last page of the section as follows:

COLLECTION

Total from | page 80

page 81

page 82

page 83

page 84

£

Total carried to Summary on page 796

## Summary

At the end of the bill, a summary is prepared for insertion of the totals from the collections of each section. To enable the total cost of any section to be established easily, it may be advisable to enter the title of each section as follows:

SUMMARY

Preliminaries | Total from | page 10

Specification preambles | page 56

Substructure | page 76

etc etc

£

TOTAL CARRIED TO FORM OF TENDER | £

Sometimes a summary is made at the end of the work sections comprising the super-structure, and the total of this is carried to a general summary at the end of the bill, thus reducing the length of the latter. At the end of the final summary, provision may be made for the addition of items such as insurance and water for works, the price of which may be dependent on the total cost of the work. All such items are described fully in the preliminaries and a note made against them that they are to be priced in the summary.

## The process of checking

When the bill has been produced, it must be checked very carefully. There are varying customs adopted to show that a page has been checked and by whom. The points to be looked for are as follows:

1) In each item:
   a) Correctness of figures.
   b) Correctness of units of measurement, especially changes from one unit to another (e.g. from cube to square or from square to linear), properly indicated.
   c) Correctness of descriptions, especially any figures contained therein, making it quite clear whether these are in metres or millimetres.
2) Generally:
   a) Sections and subsections of the bill headed properly, and headings carried over where necessary.
   b) Order of the items.
   c) Proper provision for page totals and their carrying forward or collection.

These points may appear fairly obvious, but it is extremely easy in the rush to finish a bill, which is not unusual, for some inaccuracy to be overlooked. A fascia intended to be 40 mm may be billed as 400 mm, if there has been a data entry or typing error.

## Numbering pages and items

It is important to see that the pages of the draft bill are numbered in sequence before the final print is made. When the bill is complete, one should fasten it all together and look through it, making sure that all the pages are present and in the correct sequence and that all items are referenced either by letters or by sequential numbering, the latter being best done at this stage.

## General final check

If undertaken manually, the whole of the dimensions and abstract should be examined carefully to ensure that all calculations have been ticked and all items have been lined through. Any item not dealt with should be drawn to the attention of the taker-off or other person concerned. An opportunity should be provided for the lead taker-off to look through the complete bill, as intentions may have been misinterpreted and errors in descriptions may stand out which might otherwise pass unnoticed. It should be noted that taking-off is not checked unless, perhaps, a junior is measuring for the first time. Therefore, it is advisable to make a check of the main quantities in the bill by making comparisons between items or by making approximate calculations, possibly on the following lines:

- Check the total cube of excavation with the total cube of disposal of soil.
- Calculate the total floor area of the building on all floors inside the external walls, and compare this with the total area of the floor finishes and beds.
- Make an approximate measurement of the external walls, deducting all openings, and compare with the total area of walling in the bill.
- Count the number of doors and windows, and compare with the total numbers of each in the bill. Ironmongery quantities also could be compared with the totals.
- Check the number and type of sanitary appliances with those billed.
- Measure the approximate area of roof tiling or flat roofing and check with the bill.
- Make an approximate check on any items that lend themselves to this, such as length of copings and eaves gutters, number of cupboards or other fittings, area of external pavings, and number of manholes.
- Compare the painting items with the corresponding items in other sections as far as possible; for example, take the total area of all wall or ceiling plaster compared with the areas of decoration on plaster.

These checks must not be expected to produce an exact comparison. Their principal purpose is to ensure that the quantities are not wildly out through some serious mistake; if the figures are reasonably near those in the bill, the checks will have served their purpose. Even if, owing to pressure of time, this check is left until the bill has been sent out to tenderers, errors found may be notified to them and thus taken into account.

The final bill still needs to be checked thoroughly when produced using a computerised billing system, as the different measurers may have chosen slightly different items from the library for the same piece of work, or rogues may have been prepared with slightly different wording resulting in inconsistency and duplication of items in the final bill.

## Cover and contents

The front cover of the bill should as a minimum have the title and location of the work, the date, and the surveyor's name and address. The name of the employer is also often included. A sheet should be included at the front of the bill, listing the contents; this is particularly important in the case of a large bill.

## Other bill types

Other titles for bills which have a special purpose may be encountered and are mentioned here for reference purposes. *Reduction bills* are special bills prepared when the tender figure is too high and a reduction in price is obtained by altering the work in some way.

The bill may contain omissions and additions to the original. *Addenda bills* contain details of work required which is additional to the original design, determined after completion of the original bill.

*Specialist bills* may be required to obtain tenders for specialist work, such as electrical installation, which is to form nominated subcontract work. These bills should contain the full preliminaries section of the main bill and should be accompanied by the necessary drawings.

## Bill of approximate quantities

This bill, also known as a *provisional bill*, is used when there is insufficient information available to prepare an accurate bill of quantities. Although the quantities are approximate, the descriptions should be correct. The bill is used to obtain rates for items from tenderers; as the production information becomes available, a *substitution bill* can be prepared and priced using the approximate bill as a basis.

## Schedule of prices or rates

For smaller contracts without a bill and where the specification and drawings are the contract documents, a schedule of descriptions, without quantities, can be prepared. The contractor is asked to enter rates against the items, which are then used for valuing variations to the design. The schedule seldom attempts to be comprehensive, and the rates are often unreliable. This method has also been used for contractor selection when drawn information is so scanty that even a bill of approximate quantities is impracticable.

Figures 43–45 show how the dimensions for some substructure measurement are extended and checked, and then transferred to the abstract and billed. It is important that figures can be traced from the dimensions through to the bill and that, during the final account stages, figures in a bill can be substantiated or easily amended if variations occur.

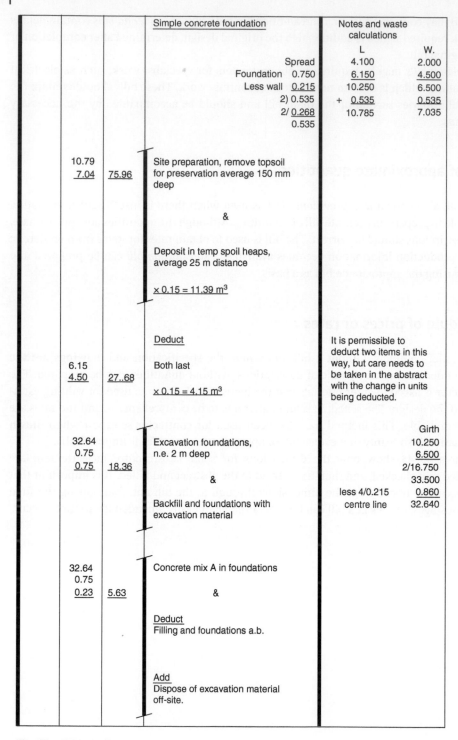

| | | | Simple concrete foundation | Notes and waste calculations | |
|---|---|---|---|---|---|
| | | | | L | W. |
| | | | Spread | 4.100 | 2.000 |
| | | | Foundation 0.750 | 6.150 | 4.500 |
| | | | Less wall 0.215 | 10.250 | 6.500 |
| | | | 2) 0.535 | + 0.535 | 0.535 |
| | | | 2/ 0.268 | 10.785 | 7.035 |
| | | | 0.535 | | |
| 10.79 | | | Site preparation, remove topsoil | | |
| 7.04 | 75.96 | | for preservation average 150 mm deep | | |
| | | | & | | |
| | | | Deposit in temp spoil heaps, average 25 m distance | | |
| | | | × 0.15 = 11.39 m³ | | |
| | | | Deduct | It is permissible to deduct two items in this way, but care needs to be taken in the abstract with the change in units being deducted. | |
| 6.15 | | | Both last | | |
| 4.50 | 27..68 | | × 0.15 = 4.15 m³ | | |
| | | | | | Girth |
| 32.64 | | | Excavation foundations, | | 10.250 |
| 0.75 | | | n.e. 2 m deep | | 6.500 |
| 0.75 | 18.36 | | | | 2/16.750 |
| | | | & | | 33.500 |
| | | | Backfill and foundations with excavation material | less 4/0.215 | 0.860 |
| | | | | centre line | 32.640 |
| 32.64 | | | Concrete mix A in foundations | | |
| 0.75 | | | | | |
| 0.23 | 5.63 | | & | | |
| | | | Deduct Filling and foundations a.b. | | |
| | | | Add Dispose of excavation material off-site. | | |

**Fig. 43** Taking-off.

| | | | | Simple concrete foundations | | | | | | |
|---|---|---|---|---|---|---|---|---|---|---|

**GROUNDWORKS**

| Site preparation | | | Excavation | | | Filling | | | Disposal | |
|---|---|---|---|---|---|---|---|---|---|---|
| Remove topsoil for preservation Average 150 mm deep | | | Excavation foundation n.e. 2 m deep | | | Filling around foundations, average thickness > 250 mm, arising from excavation | | | Disposit topsoil in temporary spoil heaps average 25 m | |
| add | | deduct | add | | deduct | add | | deduct | add | | deduct |

| Site preparation add | | deduct | Excavation add | | deduct | Filling add | | deduct | Disposal add | | deduct |
|---|---|---|---|---|---|---|---|---|---|---|---|
| 75.96 | 1 | 27.68 | 18.36 | 1 | | 18.36 | 1 | 5.63 | 11.39 | 1 | 4.15 |
| 27.68 | | 1 | | | | 5.63 | | 1 | 4.15 | | 1 |
| 48.28 | | | | | | 12.73 | | | 7.24 | | |
| | | | | | | | | | | | |
| 48 m² | | | 18 m² | | | 13 m³ | | | 7 m³ | | |

Dispose of excavation material off-site.

| add | | deduct |
|---|---|---|
| 5.63 | 1 | |
| | | |
| 6 m³ | | |

**Fig. 44** Abstract.

| | | Description of Work | Quantity | Unit | Rate | £ | p |
|---|---|---|---|---|---|---|---|
| | | GROUNDWORK | | | | | |
| A | | Site preparation, remove topsoil for preservation, average 150 mm deep | 48 | m² | | | |
| B | | Excavation, foundations, not exceeding 2 m deep | 18 | m³ | | | |
| C | | Filling to excavations, average thickness greater than 250 mm, arising from the excavations | 13 | m³ | | | |
| D | | Disposal of excavated material off-site | 6 | m³ | | | |
| E | | Disposal excavated material on-site in temporary spoil heaps not exceeding 25 m from excavation | 7 | m³ | | | |
| | | To collection | | | £ | | |

**Fig. 45**

# Appendix

# Mathematical Formulae and Applied Mensuration

## Formulae for Areas (A) of Plane Figures

Square

$$A = S \times S$$

Rectangle

$$A = L \times W$$

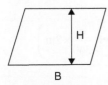

Parallelogram

$$A = B \times H$$

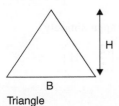

Triangle

$$A = \frac{B \times H}{2}$$

*Willis's Elements of Quantity Surveying*, Fourteenth Edition. Roy Hills and Sandra Lee.
© 2024 John Wiley & Sons Ltd. Published 2024 by John Wiley & Sons Ltd.

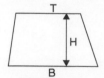

Trapezoid

$$A = \frac{(B+T)\times H}{2}$$

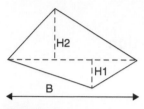

Trapezium

$$A = \frac{B+H1+B\times H2}{2}$$

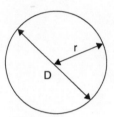

Circle

$A = \pi \times r \times r$

or

$$A = 0.7854 \times D \times D \left( \text{Note} \frac{\pi}{4} = 0.7854 \right)$$

(Circumference $= \pi \times D$ or $2\pi r$)

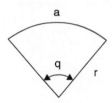

Sector of circle

$$A = \frac{r \times a}{2} \text{ or } A = \frac{q}{360}\pi r^2$$

$$\left( \text{Note length of arc} = \text{angle} \frac{q}{360} \times 2\pi r \right)$$

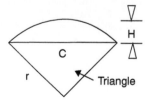

Segment of circle

$A = S - T$

Where

S = area of sector

T = area of triangle or approximately

$$A = \frac{1}{2} \times \left( \frac{H \times H \times H}{C} \right) + \left( \frac{2}{3}C \times H \right)$$

Where

H = rise

C = chord

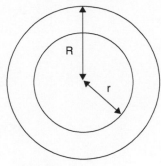

Annulus

$A = \pi(R+r) \times (R-r)$

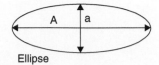

Ellipse

$A = 0.7854 \times (A \times a)$

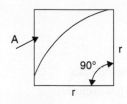

Bell mouth
(at road junction)

$A = 0.2146 \times r \times r$

## Regular Polygons

| | | |
|---|---|---|
| PENTAGON | (5 SIDES) | $A = S \times S \times 1.720$ |
| HEXAGON | (6 SIDES) | $A = S \times S \times 2.598$ |
| HEPTAGON | (7 SIDES) | $A = S \times S \times 3.634$ |
| OCTAGON | (8 SIDES) | $A = S \times S \times 4.828$ |
| NONAGON | (9 SIDES) | $A = S \times S \times 6.182$ |
| DECAGON | (10 SIDES) | $A = S \times S \times 7.694$ |

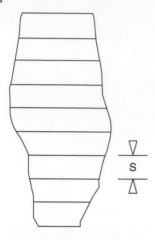

## Irregular Figures

Divide figure into trapezoids by equidistant parallel lines (ordinates or offsets)

$$A = S \times \left( \frac{P}{2} + Q \right)$$

where

    S = distance between ordinates

    P = sum of first and last ordinates

    Q = sum of intermediate ordinates

or

Simpson's rule (must be even number of trapezoids)

$$A = \frac{S}{3} \times (P + 4 \times Z + 2 \times Y)$$

where

    Z = sum of even intermediate ordinates

    Y = sum of odd intermediate ordinates

## Formulae for Surface Areas (SA) and Volume (V) of Solids

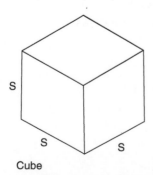

Cube

$$SA = 6 \times S \times S$$
$$V = S \times S \times S$$

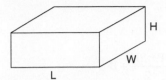

Rectangular prism

$$SA = 2(L \times W) + 2(L \times H) + 2(W \times H)$$
$$V = L \times W \times H$$

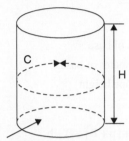

B = area of base
Cylinder

$$SA = (C \times H) + (2 \times B)$$
$$V = B \times H$$
where
  $B$ = area of base $(\pi \times r \times r)$
  $C$ = circumference $(\pi \times D)$

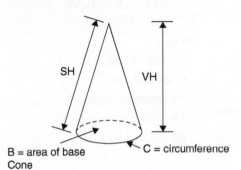

B = area of base
Cone

C = circumference

$$SA = \frac{C \times SH}{2} + B$$
$$V = \frac{B \times VH}{3}$$
where
  $B$ = area of base $(\pi \times r \times r)$
  $C$ = circumference of base
  $SH$ = slope height

b = area of top

SH    VH

R

B = area of base

Frustum of cone

$$SA = \pi \times SH \times (r + R) + b + B$$
$$V = \frac{VH}{3}\left(\pi r^2 + \pi R^2 + \sqrt{B \times b}\right)$$
where
  $B$ = area of base
  $b$ = area of top
  $R$ = radius at base
  $r$ = radius at top
  $SH$ = slope height

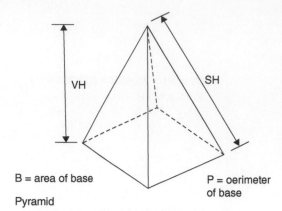

$$SA = \frac{P \times SH}{2} + B$$

(regular pyramid only)

$$V = \frac{B \times VH}{3}$$

where

    B = area of base

    P = perimeter of base

    SH = slope height

B = area of base      P = oerimeter of base

Pyramid

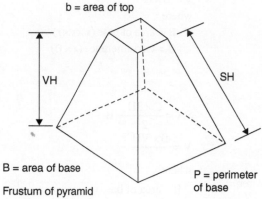

b = area of top

$$SA = \frac{SH}{2} + (p + P) + B + b$$

(regular figure only)

$$V = \frac{\left(B + b + \sqrt{B \times b}\right) \times VH}{3}$$

where

    B = area of base

    b = area of top

    P = perimeter of base

    p = perimeter of top

    SH = slope height

B = area of base      P = perimeter of base

Frustum of pyramid

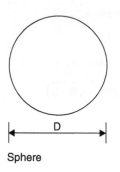

Sphere

$$SA = \pi \times D \times D$$
$$V = 0.5236 \times D \times D \times D$$

$$\left( \text{Note} \frac{\pi}{6} = 0.5236 \right)$$

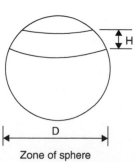

Zone of sphere

$$SA = \pi \times D \times H$$

(excluding base and top)

$$V = \frac{\pi \times H}{6} \times (3 \times R \times R + 3 \times r \times r + H \times H)$$

where

    R = radius at base

    r = radius at top

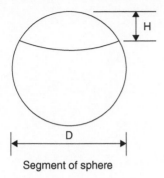

Segment of sphere

$\text{SA} = \pi \times \text{D} \times \text{H}$ (excluding base)

$$\text{V} = \frac{\pi \times \text{H}}{6} \times (3 \times \text{R} \times \text{R} + \text{H} \times \text{H})$$

where

$\text{R} = \text{radius of base}$

## Irregular Areas

Any irregular-shaped area to be measured is usually best divided into triangles, the triangles being measured individually and then added to give the area of the whole – if one of the sides, as for instance in the case of paving, is irregular or curved, the area can still be divided into triangles by the use of a compensating, or give and take, line, i.e. a line is drawn along the irregular or curved boundary in such a position that, so far as can be judged, the area of paving excluded by this line is equal to the area included beyond the boundary. In Figure A.1, the area of paving to be measured is enclosed by firm lines, the method of forming two triangles (the sum of which equals the whole area) being shown by broken lines.

For a more accurate calculation of the irregular area, particularly if evenly spaced offsets are available dividing the area into an even number of strips, Simpson's rule may be applied. The intermediate offsets should be numbered as it is necessary to distinguish the odd numbers from the even. The formula is given on page 260.

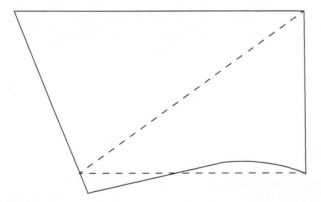

**Fig. A.1**

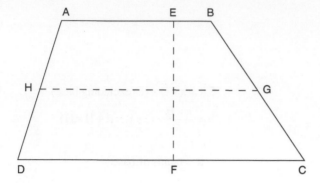

**Fig. A.2**

Where two sides of a four-sided figure are parallel to forma trapezoid, it is not necessary to divide the figure into triangles, as the area equals the length of the perpendicular between the parallel sides multiplied by the mean length between the diverging sides. In Figure A.2, the area is EF×GH, GH being drawn halfway between AB and CD and being equal to $\frac{AB+CD}{2}$, i.e. the average of AB and CD.

Another irregular figure that often puzzles the beginner is the additional area to be measured where two roads meet with the corners rounded off to a quadrant or bell mouth. This is most easily calculated as a square on the radius with a quarter circle deducted. For example, consider Figure A.3.

$$\text{Additional area} = R^2 - \frac{1}{4}\pi R^2$$
$$= R^2 - \frac{11}{4}\pi R^2$$
$$= \frac{3}{14}R^2$$

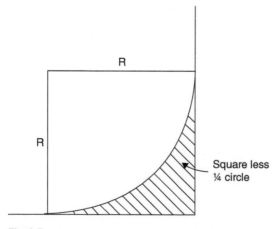

Square less
¼ circle

**Fig. A.3**

## Measurement of Arches

In the case of segmental arches, the deduction above the springing line and the girth of the arch are not usually calculated precisely, as they can be estimated sufficiently accurately, the former from a triangle with compensating lines or by taking an average height, and the latter by stepping the girth round with dividers. In the case of expensive work, one should be as accurate as possible, the measurements preferably being worked out by calculation. A rough method of measuring the area of a segment is to take 11/16 times the area of the rectangle formed by the chord and the height of the segment. This is obviously not mathematically correct, and the margin of error will vary with the radius and length of chord, but when dealing with small areas, this method will often be found sufficiently accurate. An alternative method is to take first the inscribed isosceles triangle based on the chord, leaving two much smaller segments, each of which may be scaled and set down as base $\times 2/3$ height. The error is then very small. In dealing with large areas of expensive materials, a more accurate method would be necessary, using the formula:

$$\frac{H^3}{2C}\left(\frac{2}{3}C \times H\right)$$

where C is the length of the chord and H the height.

## Excavation to Banks

The volume of excavation necessary on a level site to leave a regular sloping bank is the sectional area of the part displaced (triangle ABC in Figure A.4) multiplied by the length; the volume of the remaining excavation equals the sectional area of the rectangle BCDE also multiplied by the length. It may be, however, that the natural ground level is falling in the length of the bank, as shown in Figure A.5. The volume of earth to be displaced should then theoretically be calculated by the prismoidal formula. As the final volume is required in cubic metres, the calculation is carried out in metres:

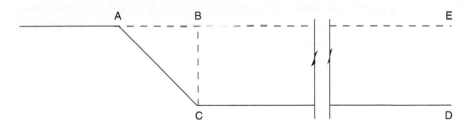

**Fig. A.4**

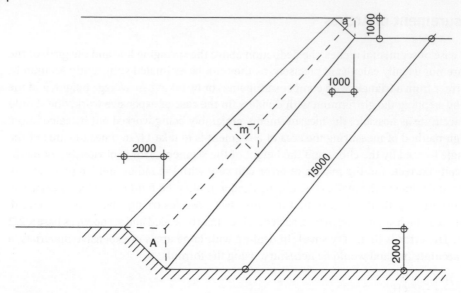

**Fig. A.5**

$$V = \frac{L(A + a + 4m)}{6}$$

where

V = Volume
L = Length
A = Sectional area at one end
a = Sectional area at the other end
m = Sectional area at the centre.

Note that in Figure A.5 the area m is not the average of areas A and a, but should be calculated from the average dimensions.

If AB and BC in Figure A.4 are both 2.00 m at the higher end and 1.00 m at the lower end, they would, assuming a regular slope, be 1.50 m at the centre. The volume of the prismoid in Figure A.5 in cubic metres would therefore be:

$$15\left(\frac{2 \times 2}{2} + \frac{1 \times 1}{2} + 4 \times \frac{1.5 \times 1.5}{2}\right)$$
$$= \frac{15(2 + 0.50 + 4.50)}{6} = \frac{105}{6}$$
$$= 17.50 \, \text{m}^3$$

In practice, however, it will often be found that so precise a calculation is not made in such cases. The surface of ground, whether level or sloping, is not like a billiard cloth, and the natural irregularities prevent the calculation from being exact. Moreover, in dealing with normal construction sites, the excavation for banks is a comparatively small proportion of the whole excavation involved (unlike, say, in the case of a railway cutting); any

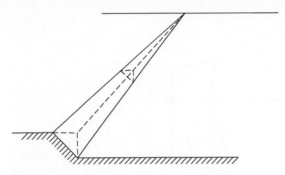

**Fig. A.6**

departure from strict mathematical accuracy due to the use of less precise methods would involve a comparatively small error. In the example given above, the volume might in practice be taken as the length multiplied by the sectional area at the centre, i.e.

$$15 \times \frac{1.50 \times 1.50}{2} = 16.88 \text{ m}^2$$

Although an error of $0.62 \text{ m}^3$ may be thought high, it must be remembered that it is being assumed that the excavation to the bank is only a small proportion of the whole.

If the natural ground level falls to such an extent that it reaches the reduced level, and the bank shown in Figure A.5 therefore dies out to nothing (as in Figure A.6), the formula will be found to simplify itself. a becomes zero. m becomes by simple geometry $\frac{A}{4}$, because the triangles at A and m are right-angled triangles, that at m having two sides enclosing the right angle each half the length of the corresponding sides of the triangle at A.

$$V = L \frac{\left( A + 0 + 4 \times \dfrac{A}{4} \right)}{6} = \frac{L \times A}{3}$$

which is the formula for the volume of a pyramid.

If the volume of earth to be excavated forms an even number of prismoids of equal length, then Simpson's rule may be applied taking the area at the offsets rather than the length as in the case of the irregular area in Figure A.5.

## Excavation to Sloping Sites

The theoretical principle for measuring the volume of excavation in cutting for a sloping bank may be extended to a sloping site (Figure A.7) If the figures given in the diagram are natural levels and it is required to excavate to a general level of +1 (the end boundaries of the area being parallel), the volume of excavation may be calculated by the prismoidal formula already given:

$$V = \frac{L(A + a + 4m)}{6}$$

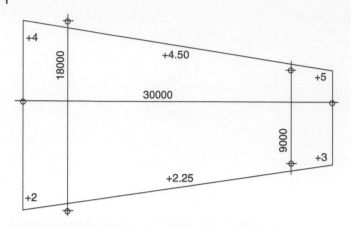

**Fig. A.7**

when

$$L = 30$$

$$A = 18 \times \frac{3+1}{2} = 36$$

$$a = 9 \times \frac{4+2}{2} = 27$$

$$m = 13.50 \times \frac{3\frac{1}{2} + 1\frac{1}{4}}{2} = 13.50 \times \frac{19}{4} \times \frac{1}{2} = \frac{256.50}{8}$$

then,

$$V = \frac{30(36+27) + \frac{256.50 \times 4}{8}}{6} = \frac{30(36+27+128.25)}{6}$$

$$= 5 \times 191.25 = 956.25 \text{ m}^3$$

If only a simply average of the four corners were taken, the result would be:

$$
\begin{array}{r}
4 \\
5 \\
3 \\
2 \\
\hline
\div 4)\overline{14}
\end{array}
$$

Average ground level   3.50

Reduced level          1.00

Average depth          2.50

$$30 \times 13.50 \times 2.50 = 1012.5 \text{ m}^3$$

This shows an error that is greater because the intermediate level on the lower boundary is not an average of the two end levels.

An average of the eight levels (giving the centre levels double value) would be:

$$
\begin{array}{rl}
& 4 \\
2/4.50 & 9 \\
& 5 \\
& 3 \\
2/2.25 & 4.50 \\
& 2 \\
\hline
& 8)\overline{27.50} \\
& 3.44 \\
& -1.00 \\
\hline
& \underline{2.44}
\end{array}
$$

or a volume of $30 \times 13.50 \times 2.44 = 988.2\,\mathrm{m}^3$, from which it will be seen that there is a definite error. This method would be satisfactory, however, if the area of the ground were a rectangle.

The error would vary with the regularity of the slope, and where the slope is fairly regular, it may often be sufficiently accurate to take an average depth over the whole area. For an ordinary site, calculation would probably be made in this way in practice.

If the formula is to be applied where end boundaries are not parallel, it will be necessary to draw parallel compensating lines along these two boundaries. It is, of course, assumed that slopes between the level given are regular. If a line of intermediate levels were given between the upper and lower lines in Figure A.7, it would be necessary to treat the two portions separately.

## Grids of Levels

On larger sites it is customary to take a grid of levels over the whole area, or at least over the area of the proposed construction, at regular intervals forming squares, plus additional levels at any significant points such as existing manhole covers or along embankments. When calculating the average level of an area covered by a grid, it is necessary to find the average level of each square of the grid by totalling the levels at the four corners and dividing by four. The average levels thus found are added together, the total is divided by the number of squares, and the result gives the average level of the area. To calculate the amount of excavation, the total area is multiplied by the difference between the average level of the area and the formation level required. If the excavation of topsoil has been measured as a separate item, then the depth of this should be deducted from the average level.

A quicker method is to total the levels in each of the categories indicated below and multiply each total by the appropriate weighting shown.

1) Levels at the external corners on the boundary of the area      Multiply by 1

2) Levels at the boundaries of the area (other than those in (1) or (3))      Multiply by 2

3) Levels at any internal corners on the boundary of the area      Multiply by 3

4) Intermediate levels within the area      Multiply 4

The results obtained are totalled and divided by a number equal to four times the number of squares (or the total number of weightings). The result is the average level of the area and will produce the same result as the first method.

## Example of Weighted Average Excavation

Assume a grid of four squares with each point 5 m apart, as shown in Figure A.8. The existing levels are included in the table and the formation that this needs to be excavated down to is at 62.000.

Existing ground levels

|  | ×1 |  | ×2 |  | ×4 |
|---|---|---|---|---|---|
| $A_1$ | 62.507 | $A_2$ | 62.750 | $B_2$ | 63.500 |
| $A_3$ | 63.100 | $B_1$ | 62.900 | | |
| $C_1$ | 63.500 | $B_3$ | 63.800 | | |
| $C_3$ | 64.100 | $C_2$ | 63.800 | | |
|  | $1 \times$ 252.950 |  | $2 \times$ 253.250 | | |
|  | | | 506.500 | | 254.000 |
|  | | | | | 506.000 |
|  | | | | | 252.950 |
|  | | | | | 16)1013.450 |

| | |
|---|---|
| Weighted average ground level = | 63.341 |
| Less formation level = | 62.000 |
| Depth of reduced level excavation = | 1.341 |

This would now be transferred to the dimension paper as follows:

| 4/ | 5.00 | Excavate to reduce levels, |
|---|---|---|
| | 5.00 | maximum depth $\leq$ 2 m |
| | 1.34 | |

Sometimes a site may be partly excavated and partly filled, and it will be necessary to plot a cut and fill contour using interpolation, as shown below, to ascertain the location. Contours may also be plotted to represent the division between depth bands of excavation

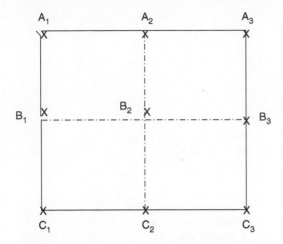

**Fig. A.8**

or fill as required by the Standard Method of Measurement. Grid squares that are cut by the contour line to form triangles or trapezia should be dealt with separately; the average level of each is found by adding the levels at the corners and dividing by the number of corners. The area of each irregular figure is multiplied by the difference between its average level and the formation level to give the volume of excavation or fill to add to the main quantity. When measuring excavation work, one should always look for sudden changes in levels which may indicate items such as embankments or craters. These areas should be dealt with separately and not averaged within the main area as above.

## Interpolation of Levels

When measuring excavation work, it is sometimes necessary to ascertain the ground level at a point between two levels given on the drawing or to locate a point at which a certain level occurs. This can be achieved by interpolation as shown in the following:

Give two levels at points A and B 50 m apart, find an intermediate level at point C 20 m from A.

Level at A = 3.60
Level at B = 2.90
Difference 0.70
Therefore, the ground falls 0.70 over 50 m
To find the fall y over 20 m

$$\frac{0.70}{50} = \frac{y}{20}$$
$$20 \times 0.70 = 50y$$
$$y = 0.28$$

The level at C = 3.60 − 0.28 = 3.32

If the intermediate level is known, such as a formation level, then the distance from A could be found in a similar manner, the unknown being the distance rather than the level.

# Index

*Willis's Elements of Quantity Surveying,* Fourteenth Edition. Roy Hills and Sandra Lee.
© 2024 John Wiley & Sons Ltd. Published 2024 by John Wiley & Sons Ltd.